CLIMATE SENSE

D.M.MACFARLANE BSc. MSc.

Author: Macfarlane, D. M.

Title: CLIMATE SENSE: A Layman's Guide to Climate Change
ISBN: 9781794452770

Imprint: Independently published

ABOUT THE AUTHOR

David Macfarlane is a Sydney-based entrepreneur /building designer with a BSc in environmental architecture (UNSW, 1993) and a Masters in sustainable building design (Oxford Brookes, 2013). Amongst other projects, David designed, built, and operated for fourteen years a ten-cabin eco lodge on a remote national-park island on the Great Barrier Reef. His most recent project is northern Tasmania's Tamar Solarhome (*www.tamarsolarhome.com*), an all-electric, totally off-grid house.

PRAISE FOR *CLIMATE SENSE*

"I loved your book. You seem to have successfully avoided the technical issues that befuddle non-scientists and have relied on logic and common sense (that also, alas, befuddle part of our elite, but are nonetheless more accessible to those with somewhat open minds)." Dr Richard Lindzen (American atmospheric physicist and a lead author of the IPCC's Third Assessment Report on climate change).

"Thanks for sending your book, it is very well done and covers some different territory than what I have seen before." Dr Judith Curry (American climatologist and former chair of the School of Earth and Atmospheric Sciences at the Georgia Institute of Technology).

"Ideologies are simple ideas, disguised as science or philosophy, that purport to explain the complexity of the world and offer remedies that will perfect it. Ideologues are people who pretend they know how to "make the world a better place" before they've taken care of their own chaos within. The warrior identity that their ideology gives them covers over that chaos. Ideologies are substitutes for true knowledge, and ideologues are always dangerous when they come to power, because a simple-minded I-know-it-all approach is no match for the complexity of existence. Furthermore, when their social contraptions fail to fly, ideologues blame not themselves but all who see through the simplifications."

Dr. Norman Doidge
(author of *The Brain That Changes Itself*)

"Those who are governed by reason desire nothing for themselves which they do not also desire for the rest of humankind."

Baruch Spinoza

"Socrates used to call the popular beliefs 'bogies', things to frighten children with."

Marcus Aurelius; Meditations

For humanity and our environment

CONTENTS

PREFACE

I make no claims of expertise in regards to climate science. I'm simply attempting to understand and explain in layman's terms what many scientists are saying about climate change – experienced, well-credentialed scientists that for a variety of reasons (which I also explore) are not being acknowledged by the authorities or mainstream media.

I have not included graphs and charts in this book because I discovered that scientists can (and, in fact, usually do) manipulate data and scales in order to create graphs that support their own particular bias; to the extent that they almost become meaningless. Nor have I specifically referenced observations I've made from viewing online presentations, mainly because I hope to encourage readers to watch and listen to at least some of the debates, lectures, and interviews with the scientists I've listed and form their own opinion instead of taking my word for it. And most importantly, I hope readers will be motivated to get better informed and think deeply about this important issue rather than simply following what *appears* to be the general consensus. In other words, if you don't believe or trust what I've written, go directly to the source and find out for yourselves.

When I showed first drafts of this book to friends, family, and acquaintances, the most common reaction from those most determined not to question their own deeply-held beliefs – after, I suspect, reading a couple of chapters at most – was nearly always *"well that's just your opinion"*. And that's when I realised what I was up against. Apparently, many people, no matter how intelligent and well-educated, either can't tell the difference between opinion and observation, or simply use that claim to dismiss information that for some reason they don't want to hear. Either way, they're certainly not interested in challenging their hard-held beliefs.

If I told someone I'd just seen an elephant in their backyard, and they replied *"well that's just your opinion"*, I'd reply; *"no, it's not, it's my observation"*. Of course an elephant in anyone's backyard would seem so unlikely that they may question my eyesight, my motives, and even my sanity. But if I asked them to have a quick look for themselves, and they refused, I'd wonder at *their* motives and sanity, and their lack of curiosity. Why would they not bother to look? Or perhaps more significantly, why are they afraid to look? Are they afraid of what they might find? Perhaps it's just not socially acceptable to have an elephant in your backyard?

They might insist that they care deeply about what *is* in their backyard, and have checked with all their neighbours, the local council, the TV and radio stations, and many prominent experts, who all assure them there are no elephants to be seen. So they fervently believe there are none in their backyard; because they trust and believe the general consensus and the authorities, who, as we all know, always tell the truth and are always right.

But still they won't look for themselves.

It's a relevant analogy because many months after reading part of my first draft, as far as I'm aware, none of those who declared *"it's just your opinion"* have watched the easily-accessible online lectures by any of the climate scientists I've referenced. For some reason they don't want to look. And that's dangerous. Because uninformed or misinformed people, with fixed opinions and beliefs, who are too frightened or too indifferent to learn something new, are far too easily deceived and led by those with ulterior motives.

I make a lot of effort in this book to separate observation from opinion. Although obviously I've had to make judgements on what observations to include, and my conclusions are my opinions based on those observations.

However, when I write, for instance, that a senior NASA scientist says something in particular about climate change, that is my observation, not my opinion. In fact, it is the opinion of a highly qualified scientist worth listening to; supported by statistical evidence and many years of professional research.

Despite having read, watched, and believed hundreds of books, articles, documentaries, and presentations about global warming – or as it's now called, 'climate change' – over the last thirty years, it wasn't until I started researching and writing this book that I realised how little I actually knew about something so vitally important to my own career.

Caring for the environment has been, and still is, the driving force behind most of my major decisions. But, as I was soon to discover – in regards to the science and politics of climate change – it wasn't that I was uninformed, more like I was just badly misinformed.

So all I ask is that you read this whole book with an open mind, follow the reasoning, double check the facts, watch and listen to some of the scientists I reference, and see what conclusion *you* come to. It may surprise you to discover that it's not the same as that peddled with such factual and moral certainty by the majority of mainstream media, celebrities, academia, and government authorities. And if nothing else, you'll learn a lot of things you probably have never heard before. For the first time in your life you may start to question the general narrative we've all been led to believe for the past fifty years – that humans are destroying the planet.

The key message of this book is that mainstream media is not presenting us with all the facts in an unbiased manner on many issues, particularly in regards to climate science. And the purpose of the book is to encourage people to do a little research before committing themselves to a belief system that could fundamentally change the world.

Climate change has always been (and still is) presented as an argument between scientists and science 'deniers'. But in reality, it can more honestly be presented as an argument between 'alarmist' scientists and 'non-alarmist' scientists.

I was shocked to discover that there are so many well-credentialed, experienced, credible scientists who are not climate change alarmists, and whose arguments (to me at least) actually sound more reasonable and scientific than the (often) emotive arguments of alarmists. In fact it was quite easy to find out that the constantly-made claim that 97% of scientists are climate alarmists is far from the truth.

The first question this raises is; why had I never heard of these non-alarmist scientists before? And the answer appears to be that there are at least two forms of 'censorship' currently operating. While speaking out or writing anything against the accepted narrative of climate change alarmism isn't actually illegal, or officially censored, it does suffer from what is known on the internet as 'shadow banning'. Articles and presentations can very easily be made very hard to find.

This appears to have happened because journalists and publishers, in general, have been taught in the past few decades to be advocates rather than reporters. And as such, tend to promote social justice issues that support their particular belief system. They try to prove their beliefs rather than explore issues and discover the truth.

All the free-to-air TV stations and all the major newspapers bar one in Australia currently present climate change alarmism as a matter of scientific fact. As far as I'm aware, they have never given any exposure to any of the non-alarmist scientists I discovered online.

The other form of censorship prevalent at the moment is the 'self-censorship' of social acceptance when an issue is framed as a moral issue, such as caring for the environment or saving the planet, which is virtually impossible to argue against.

When I decided to start exploring the 'other side' of the climate change debate, I searched my local bookshop. On the bottom shelf of a wall of books promoting the climate change crisis I found one lonely book by an Australian scientist critical of climate change alarmism. The young shop assistant sneered at me silently when I paid for it, and I sat for the next hour in a dark corner of the café reading the book with the cover folded back so no one could see the 'immoral garbage' I was reading. That's a particularly effective form of censorship, regardless of whether or not it was a product of my own imagination. And it's a serious impediment to discovering and disseminating the truth, especially with such a complex issue.

Despite the negative consequences of delving into an issue that has only one socially acceptable narrative, the reason I've written this book is because I feel it's my responsibility to pass on the things I've discovered. Not many people have the motivation or opportunity to do such in-depth research, and very few people have the thirty-plus years of practical experience working with off-grid power systems that I've had. So I believe that my observations may have some value.

Am I absolutely sure I'm right in the conclusion I've come to? Of course not, it's just my opinion, and I'll change it if new evidence convinces me otherwise. But five years ago, after hearing only one side of the argument for more than twenty years, I was about 90% sure that we were facing a climate crisis unless we stopped burning fossil fuels. Now, after becoming far better informed by listening to both sides of the argument, plus discovering the true limitations of so-called renewable energy, I'm about 90% sure that there is no such crisis.

Perhaps the real crisis is the corruption of modern science, journalism, and academia, which in the last thirty years has propagated the unchallengeable belief that CO_2 (an invisible odourless gas that is vital for plant growth) is a dangerous pollutant that controls the earth's climate.

While most scientists appear to agree that the earth has been warming at a rate of about 1°C per century since the end of the last mini ice age two hundred years ago (with a corresponding rise in sea level of around 12cm per century), alarmists claim, with very little evidence apart from their own computer modelling, that this warming has suddenly increased recently because of the fossil fuels being burnt to power the industrial revolution. The trouble is, they decided this was a problem before it happened. And then it didn't.

Fossil fuel consumption and CO_2 emissions have doubled in the last thirty years, without a corresponding rise in sea level, and for half of that period there was no warming. So alarmists changed their focus, claiming that CO_2 emissions were causing extreme weather. But this is a claim constantly made by mainstream media (and insurance companies), not by scientists with evidence. Not even the IPCC (the main UN authority promoting climate change alarmism) claims to have solid evidence that there has been a recent increase in extreme weather.

Trying to reduce CO_2 emissions, which if acted on seriously (by eliminating the use of all fossil fuels) will inevitably destroy many country's economies and standard of living. It will do nothing to help the environment, or 'save the planet'.

Preventing and cleaning up air, water, and land pollution, disposal of waste efficiently and cleanly, preserving natural ecosystems, managing forests, and preventing bushfires all cost money; and are usually the first things to suffer when an economy is struggling. It's no coincidence that the poorest countries in the world are nearly always the countries with the most damaged environments.

After what I've recently discovered, I strongly suspect that, at its core, the modern climate change movement has no genuine concern for the environment or humanity. It's a deceitful, nihilistic political movement largely followed by well-meaning but misinformed idealists and a general public who innocently believe what they're told by mainstream media. If allowed to proceed unabated, it will destroy modern civilization, and humanity.

I know this is hard to believe for anyone who has been fed a steady diet of environmentalism and climate change alarmism for the past twenty years. But read on, you may find yourself convinced.

1. AN ECHO CHAMBER

"A metaphorical description of a situation in which beliefs are amplified or reinforced by communication and repetition inside a closed system."
Wikipedia

Up until recently, I was quite certain I knew enough about global warming and climate change to have a strong opinion. I'd seen the graphs that showed the temperature and carbon dioxide emissions and sea levels all rising together, and others that showed the catastrophic projections.

I'd seen the pictures and diagrams of wild storms, floods, droughts, and wildfires; sea levels rising, islands disappearing, coastlines eroding away, ice caps melting, glaciers receding, coral reefs bleaching at an alarming rate, and polar bears dying. And it's killing an estimated 315,000 people every year. Apparently, climate change is causing more disease, more famine, and more war, creating hundreds of millions of refugees and causing the destruction of entire ecosystems and species.

It seemed obvious that climate change is real, it's caused by humans burning fossil fuels, it's going to destroy the planet and kill billions of people, and something has to be done about it urgently before we reach an irreversible tipping point. The science is settled; and 97% of scientists agree.

I didn't think I needed to know much more than that, and there didn't seem to be much point reading *another* book on the catastrophe of climate change. It was all too depressing.

About six months ago it suddenly occurred to me that I didn't really know much at all about the science of climate change; except that 97% of scientists apparently agreed. But like most people with a strong opinion about climate change, I couldn't name one climate scientist. I'd never actually listened to a group of real scientists debating or explaining the issues.

In fact, everything I'd read and heard about climate change began with the assumption that all sensible well-informed people agreed that it was caused by humans burning fossil fuels. That was never really questioned. But I now realised that although Al Gore, Malcolm Turnbull, Barack Obama, and even Tim Flannery might be smart, articulate guys, they're not climate scientists.

And it got me thinking.

Several years ago, I attended a presentation by a motivational speaker. He started by saying that he was first going to test our levels of concentration by showing a short video clip, explaining that it was of a basketball team, and he wanted us to count how many times they passed the ball.

When it finished he asked how many of us counted twenty passes. Everyone put their hand up, it hadn't been that difficult. Then he asked if anyone had noticed anything unusual. A few people put their hands up, and he asked if they had seen the video before, which they all admitted they had.

"Good", he said, *"then you'll be able to confirm that when I play the clip again it's exactly the same one"*.

When he played it again, as the players moved around passing the ball to each other, much to everyone's astonishment a guy in a gorilla suit walked across to the front and centre of the court and stood waving his arms around for ten seconds before walking off. How could we possibly not have noticed? It's a pertinent lesson I always try to remember. We often only see what we're looking for, what we're focused on, or what we want to see. And we can completely fail to see something blatantly obvious, right in front of our nose.

As philosopher Edward de Bono puts it: *"Most executives, many scientists, and almost all business school graduates believe that if you analyse data, this will give you new ideas. Unfortunately, this belief is totally wrong. The mind can only see what it is prepared to see."*

I also started to think about where we get our information from, and how things have changed in that regard over the course of history.

In the mid-fifteenth century, the printing press was invented. Since the first book printed in volume was the Bible, it had a profound effect on Christianity. One of the effects was that people who had learnt to read could go directly to the source and read the words of Christ as quoted in the New Testament, instead of having to rely on the interpretation of a group of others who could easily control and manipulate their beliefs. As a result, the Catholic Church gradually lost power. People were one step closer to Christ's actual teachings.

But it was still second-hand information, at the discretion of the witness's interpretations. Imagine how much more truthful, powerful, and enlightening it would have been if people had video recordings of Christ preaching, where you could hear his actual words, the sincerity in his voice, and see the passion in his eyes.

In recent years, another revolution in information dissemination has taken place, perhaps just as profound as the printing press. And it's *not*, as you might think, TV, radio, websites, Twitter, Instagram, Facebook, or even Wikipedia.

Many people would probably agree that mainstream media have been dumbing things down for years, under the self-fulfilling assumption that most of us have a short attention span, aren't particularly smart, and just want to be told what to think. In recent years social media has put that trend on steroids. There's very little opinion-free news, almost no genuine open-minded analysis, and every outlet has an agenda (both political and sensationalist), resulting in heavily-edited sound bites of click-bait-chasing infotainment masquerading as serious news. Thinking you're getting a balanced view by watching and reading a variety of channels and newspapers is like thinking you've got a balanced diet by alternating between McDonalds, KFC, and Pizza Hut.

So these days, if you get all your information from mainstream/social media, you're not just uninformed, you're misinformed, and there's not much you can do about it, no matter how smart you think you are.

Until you discover the incredible power of the internet For the first time in human history we can access unedited, unfiltered information, direct from the source. So instead of having to depend on someone's opinion of what someone else has said, or their carefully-edited sound bites taken out of context to satisfy a narrative, we can watch qualified experts speaking, unedited, in their own words, in context.

And, if you're curious, have an open mind, and listen attentively, you may even be able to form your own opinion instead of parroting the opinions of your favourite pundit on mainstream media.

It does require some discretion to navigate your way around the internet, but when you can see and hear people talk, it's not that difficult to tell the delusional fanatics from the sincere well-reasoned thinkers with relevant qualifications and knowledge. And of course, these days it's easy to check someone's credentials.

Long-form interviews, lectures, and presentations, many of them two-or-three hours duration that have been viewed hundreds of thousands of times (they all display a view-counter) belie mainstream media's low opinion of the general public's attention span, level of interest, and intelligence. It must shock TV executives who bother to discover that there are more people viewing some of these presentations online than are watching their primetime TV programs.

You can watch five-hour-long US senate enquires where panels of experienced scientists try their best to explain complex issues to clueless politicians. You can also watch actual meetings between international presidents, live-streamed direct from the Oval Office, which, interestingly, are nothing like what is generally reported in mainstream media.

I've given up on TV, radio, and newspapers as a credible source of information, and instead have been able to watch hundreds of hours of interviews, lectures, and presentations by well-credentialed scientists, social psychologists, and philosophers from some of the most respected universities and government institutions in the world.

Of course, when there are apparently credible experts presenting completely opposite conclusions, we have to listen to both sides with an open mind and use our own judgement to decide who makes more sense, sounds more logical and reasonable, and appears to have more integrity.

And we have to be wary that we don't decide what we believe in first then listen only to people who confirm our opinion; which might make us feel more comfortable, perhaps even slightly smug, however it won't lead to the truth.

So, keeping these things in mind, and setting aside my own prejudices and beliefs as much as humanly possible, I set out on a mission to find out all I could about the science, philosophy, and politics behind the climate change movement.

It turned out to be a fascinating, but challenging, journey.

2. THE CONSENSUS

If you tell a big enough lie and tell it
frequently enough, it will be believed.
*"In the big lie there is always a certain force of
credibility; because the broad masses of a nation are
always more easily corrupted in the deeper strata of
their emotional nature than consciously or
voluntarily; and thus in the primitive simplicity of
their minds they more readily fall victims to the big lie
than the small lie."*
Adolf Hitler (Mein Kampf)

The first thing I discovered, much to my surprise, was that the oft-made claim that 97% of scientists agree that humans are a major driver of climate change may not be quite right. In fact, a quick search of YouTube reveals dozens of scientists claiming just the opposite. Of course these must be the 'deniers' I've heard of that make up a tiny group of climate-change sceptics; angry right-wing conspiracy theorists who have sold their souls to the fossil fuel industry for nothing other than personal profit.

The first I listened to was Dr Patrick Moore, surprisingly, a founding member and former president of Greenpeace with a PhD in Ecology. Apparently, he left Greenpeace in 1986 after it had morphed from what he describes as a humanist environmental movement into an anti-humanist Marxist-inspired political movement. That's an outrageous claim, but he seems quite rational for a 'denier'. Obviously he must have lost the plot somewhere along the line. Perhaps he was being sponsored by a fossil-fuel company.

But then I discovered Dr Roy Spencer, a former Principal Research Scientist at the University of Alabama and Senior Scientist for Climate Studies at NASA's Marshall Space Flight Centre. Surely a NASA scientist couldn't be a 'denier'? Once again, he seems surprisingly rational. Dr Spencer's website describes his views on climate change:

"This website describes evidence from my group's government-funded research that suggests global warming is mostly natural, and that the climate system is quite insensitive to humanity's greenhouse gas emissions and aerosol pollution. Believe it or not, very little research has ever been funded to search for natural mechanisms of warming – it has simply been assumed that global warming is manmade. This assumption is rather easy for scientists since we do not have enough accurate global data for a long enough period of time to see whether there are natural warming mechanisms at work. ... This is

actually quite easy for meteorologists to believe, since we understand how complex weather processes are. Your local TV meteorologist is probably a closet skeptic regarding mankind's influence on climate. Climate change – it happens, with or without our help." [www.drroyspencer.com]

Okay, so I've discovered a couple of scientists who disagree with the general consensus. Big deal. As we're constantly being assured, a few rogues amongst thousands of scientists doesn't prove anything.

But then there's German Physicist and meteorologist Klaus-Eckart, who states that: *"Ten years ago I simply parroted what the IPCC (Intergovernmental Panel on Climate Change) told us. One day I started checking the facts and data – first I started with a sense of doubt but then I became outraged when I discovered that much of what the IPCC and the media were telling us was sheer nonsense and was not even supported by any scientific facts and measurements. To this day I still feel shame that as a scientist I made presentations of their science without first checking it ... scientifically it is sheer absurdity to think we can get a nice climate by turning a CO_2 adjustment knob."* [Translated from a 2012 interview in Swiss magazine *Factum*]

And University of California's Professor Hal Lewis's letter of resignation to the American Physical Society in October 2010:

"For reasons that will soon become clear my former pride at being an APS Fellow all these years has been turned into shame, and I am forced, with no pleasure at all, to offer you my resignation from the Society. It is of course, the global warming scam, with the (literally) trillions of dollars driving it, that has corrupted so many scientists, and has carried APS before it like a rogue wave. It is the greatest and most successful pseudoscientific fraud I have seen in my long life as a physicist .... I want no part of it, so please accept my resignation." [www.heartland.org]

University of Melbourne Professor Ian Plimer puts it just as definitively, claiming that 'climate science' is not even a legitimate field of science, but is an aggregation of many different and disparate scientific disciplines:

"It is an invention of an exclusive club to exclude all those mathematicians, physicists, chemists, biologists, astronomers, geologists, and meteorologists who don't follow the ideology ... a clique, mainly comprising computer modellers and meteorologists, (which) excludes those with the most

to offer such as solar physicists, astronomers, geologists, and carbon dioxide chemists."
[*Climate Change Delusion and The Great Electricity Rip-off*]

By now I'm feeling quite shocked. And the more I looked, the more 'deniers' I found, including the following scientists that can all be viewed on YouTube giving full-length, unedited lectures, presentations, and interviews. They include:

Dr Don Easterbrook (Professor Emeritus at Western Washington University)

Dr Jennifer Marohasy (Australian biologist and former director of the Australian Environment Foundation)

Dr Bob Carter (former professor and head of the School of Earth Sciences at James Cook University)

Dr Peter Ridd (Australian marine physicist formerly with James Cook University)

Dr John Christy (climate scientist at the University of Alabama)

Dr William Happer (Professor of Physics at Princeton University)

Dr Freeman Dyson (professor emeritus in the Institute for Advanced Study in Princeton)

Dr Nir Shaviv (professor at the Racah Institute of Physics of the Hebrew University of Jerusalem)

Dr Fred Goldberg (Royal Institute of Technology)

Dr Judith Curry (an American climatologist and former chair of the School of Earth and Atmospheric Sciences at the Georgia Institute of Technology)

Dr Timothy Patterson (a professor at Carleton University Ontario specialising in paleoceanography and paleoclimatology)

Dr Richard Lindzen (Professor of Meteorology at the Massachusetts Institute of Technology)

Dr Tim Ball (former professor in the Department of Geography at the University of Winnipeg).

All of these experienced scientists appear to be well-reasoned and sensible, and are obviously well-credentialed. A few of them admit they have done consultancy work in the past for fossil fuel companies but claim they could make far more money and be far more popular if they instead pushed the alarmist agenda. Some declare that the large majority of 'deniers' have never received a cent from any private organisation. And many of them claim to have been bullied, vilified, and in some cases forced out of their academic positions on spurious charges for daring to question the general 'consensus'.

Like I've said earlier, don't take my word for it, or for that matter, the word of the 'experts' in newspapers and on TV telling you what to think. Go online, search for any of the above scientists, listen and watch carefully, and decide for yourselves. Like me, you may find them credible and convincing.

And you'll soon discover that they are certainly not on their own, or scientific outcasts. In fact, over the past couple of decades, there have been dozens of petitions and appeals signed by literally thousands of scientists denouncing the climate change alarmism promoted by the IPCC and mainstream media. All of these petitions have been denigrated and dismissed by alarmist organisations, backed up by most of mainstream media.

The main petitions of scientists opposing the climate change alarmism promoted by the IPCC include:

The Heidelberg Appeal (4000 signatures including 62 Nobel prize winners)
The Oregon Petition (31,000 accredited scientists)
The Manhattan Declaration (600 research climatologists)
The Petition to the United Nations (100 geoscientists)
Petition to the Canadian Prime Minister (60 climate experts)
The Leipzig Declaration (100 geoscientists)
The Statement from Atmospheric Scientists (50)

Petition to the German Chancellor (200 German scientists)

Statement from the American Physical Society (150 physical scientists)

Petition to President Obama (100 leading climate researchers)

UN Climate Scientists speak out on Global Warming (700, many previously involved with the IPCC)

All of these petitions are critical of global warming alarmism, and all of them (with signatures and accreditations) are accessible via Google.

The Heidelberg Appeal and Oregon Petition.

"This historical document signed by more than 4,000 distinguished scientists, including 70 Nobel Laureates, was released in the beginning of the infamous "Earth Summit" (Rio de Janeiro, 1992) to oppose environmental obscurantism, including climate alarmism. Among other things, the Heidelberg Appeal said:

'We want to make our full contribution to the preservation of our common heritage, the Earth. We are, however, worried at the dawn of the twenty-first century, at the emergence of an irrational ideology which is opposed to scientific and industrial progress and impedes economic and social development. We contend that a Natural State, sometimes idealized by

movements with a tendency to look toward the past, does not exist and has probably never existed. We intend to assert science's responsibility and duties toward society as a whole. We do, however, forewarn the authorities in charge of our planet's destiny against decisions which are supported by pseudoscientific arguments of false and non-relevant data. The greatest evils which stalk our Earth are ignorance and oppression, and not Science, Technology, and Industry.'
These and other appeals, petitions, and declarations by distinguished scientists against climate alarmism have been largely ignored by the formerly mainstream media and buried in slander by alarmist front groups and hard Left mouthpieces."

[www.defyccc.com/heidelberg-appeal-anniversary]

According to the 'deniers' you have to be careful where you get your information from. Wikipedia, for example is apparently policed by the pro-climate-change-alarmism 'community'. All of its pages on climate change state as a matter of fact that it is caused by human-induced CO_2 emissions.

Experiments have been done by 'deniers' where they have edited sentences to state that some scientists disagree, and these changes have disappeared within minutes. And scientists that are being vilified for their dissenting opinions on climate change claim they've given up trying to correct the lies and

43

misinformation on their own Wikipedia pages. While that may sound conspiratorial, and it may not still be the case, but apparently at one stage the main Wikipedia editors were leading members of the Green party. It's also obvious that Google are using algorithms to push politically correct (pro-alarmist) web pages to the fore.

So there's no doubt that there are many respected scientists who strongly disagree with climate change alarmism. But where did the consensus figure come from?

Following is a scientist's analysis of the original paper (by Cook et al, 2013) that started the 97% consensus myth which is so often quoted by climate change alarmists:

"The consensus Cook considered was the standard definition: that Man had caused most post-1950 warming. … the true consensus among published scientific papers is now demonstrated to be not 97.1%, as Cook had claimed, but only 0.3%. Only 41 out of the 11,944 published climate papers Cook examined explicitly stated that Man caused most of the warming since 1950. Also, he arbitrarily excluded about 8000 of the 12,000 papers in his sample on the

unacceptable ground that they had expressed no opinion on the climate consensus. These artifices let him reach the unjustifiable conclusion that there was a 97.1% consensus when there was not. In fact, Cook's paper provides the clearest available statistical evidence that there is scarcely any explicit support among scientists for the consensus that the IPCC, politicians, bureaucrats, academics and the media have so long and so falsely proclaimed."
[www. wattsupwiththat.com]

When a climate alarmist makes the claim that 97% of scientists agree with them, it would be reasonable to ask the following four questions:

1. What was the specific survey showing this result?

2. How many scientists were questioned in that particular survey?

3. How were those scientists selected?

4. What was the precise wording of the survey question?

Because as always, especially with statistics and surveys, the devil is in the detail. And surveys are particularly easy to manipulate. For example, if it was generally accepted that clearing a large area of forest would most likely have some effect on the local climate, you could hypothetically argue that felling one tree will also have some effect on the world's climate. But to what degree? How significant would it be? Should you then ban the felling of any trees at all in order to 'save the planet'?

There are many aspects of human activity over the last hundred years (including massive changes in land use, deforestation, a human population increase from one billion to seven billion, and of course many tonnes of fossil fuel burnt to power the industrial revolution) that undoubtedly have had an effect on the climate. So you're likely to get at least 97% consensus when you ask if humans are having *some* effect on the climate. But again, to what degree?

There are several surveys claiming to prove consensus, but the only ones that get anywhere close to 100% are when the question uses the word *'some'*, which as I've indicated, from a strictly scientific point of view is virtually impossible to disagree with. However, it's a huge jump to then conclude that: *therefore we're in danger of catastrophic climate change unless we stop using fossil fuels.*

So not only have alarmists grossly exaggerated a perceived problem, but they have also jumped straight to a solution with no analysis. It looks suspiciously like they've started with the answer they want (to stop burning fossil fuels), and then worked backwards from there, no matter how hard it was to join the dots. If nothing else, it demonstrates a remarkable lack of scientific integrity.

This, on its own, puts a huge hole in the credibility of the alarmist's argument. An argument I always used myself when asked my opinion about climate change. If 97% of scientists agreed there was a serious problem we had to act on urgently, that was good enough for me. I didn't really need to know any more than that. I could maintain that opinion by deliberately avoiding any in-depth inquiry to become better informed.   Then I could adopt a 'belief', doggedly defend it, and demonise anyone who dared to disagree … which sounds familiar.

The fact is, there is nowhere near a 97% consensus. Which means, of course, that there have been a lot of people in positions of authority and influence telling us, either deliberately or inadvertently, a blatant lie for many years.

The question is, why?

And why have most of us believed it so readily – otherwise well-informed people, including myself, who haven't bothered to do even the most basic research for such an important issue? Why have I never looked seriously, with an open mind, at the other side of this 'climate change' argument?

Probably because I thought (perhaps a little arrogantly) that I was so well-informed. And I had no reason to doubt the assertions and opinions of the respected environmentalists and scientists whose books I'd read, or the mainstream media who predominantly support their beliefs. It all seemed pretty obvious anyway. We were the good guys, who really cared about the environment, and anyone who didn't agree with us was a greedy, science-denying capitalist (undoubtedly working for a fossil fuel company) who cared nothing about the future of the planet. Simple really.

That's about all you need to know about climate change. The science is settled. We know this because mainstream media and politicians and actors and celebrities and artists and school children tell us so. And we know that all decent, responsible, caring, well-informed people believe that human-induced climate change is real. It's obvious.

Or so I thought.

3. SOMETHING SHIFTY

"The problem with the world is that
fools and fanatics are so certain of themselves,
and wiser people are full of doubts."
Bertrand Russell

I have to admit that in recent years there were things being said and done in support of the 'climate-change crisis' that had made me somewhat uneasy.

Climate change believers always appear to be very certain they are right and seem to be trying to shut down debate or dissent by asserting that 'the science is settled'. That attitude always worried me. I like to hear opposition to my opinions and beliefs – it either helps me strengthen my arguments or forces me to rethink them.

Following is the sort of response that has become common when someone dares to question climate change certainty:

"When we put it to air, people were terribly angry at us for airing such extremist views, especially from environmental organisations that were annoyed that we had given the stage to these kinds of views."
[Israel Public Broadcasting Corporation, December 15, 2019]

And here are the 'extremist views' which elicited such censure:

"The last thing we should do is drive our industries offshore and be putting pressure on household budgets and risk third world-style blackouts all in the name of climate change. We have got to be sensible and balanced and proportionate about these things and I don't think other policy makers are right now."
[Former Australian Prime Minister Tony Abbott]

Over the years, I've watched dozens of TV panel shows where a group of climate change 'experts', backed up by a supportive host and compliant audience, have denigrated and mocked a token climate change 'denier' who dares to question their 'scientific consensus'. More often than not, they choose (perhaps deliberately) a 'denier' who's not that well-informed and is easy to portray as a crackpot. It always seems like a setup, and there's never a serious attempt to understand their argument.

Any good scientist will tell you that true science is the constant process of questioning and testing hypotheses. It's rarely if ever 'settled', especially with something as variable and complex as the earth's climate. And consensus, even if true, is the realm of politics, not science.

A further indication that something wasn't quite right was the devious and sometimes telling use of language, which to me, reeks of subtle manipulation. Labelling people as 'deniers' (with obvious implications of a moral deficit shared with holocaust deniers) if they didn't agree with the consensus, is a sly and effective way of silencing dissenters.

When it was discovered that there was a pause in global temperature increases after 1998 [discussed in detail later] the issue mysteriously changed from 'global warming' to 'climate change', which of course is impossible to disagree with, but a meaningless claim. Plus the introduction of the word 'belief' into the debate was disconcerting (as in "do you believe in climate change"). Belief is a word used in religion, not science.

Then, of course, there was the infamous 'Climategate' affair of 2009, when thousands of emails were released by an anonymous whistle-blower, revealing communication between the main scientists driving the IPCC's alarmist agenda, which indicated a definite intent to manipulate data, silence dissenters, and control the peer-review process.

One of the most damning was lead scientist Phil Jones to his colleagues in November 1999:

"I've just completed Mike's Nature trick of adding in the real temperatures to each series for the last 20 years (ie. from 1981 onwards) and from 1961 for Keith's to hide the decline."
[www.undeceivingourselves.org/I-ipcc.htm]

The other thing I began to realise was that nearly all of the information supporting action on climate change is emotive, fearful, alarmist, and apocalyptic. That's never a good way to figure out the truth, or make wise decisions. It's certainly not scientific. We lose our ability to think rationally when we're frightened. And there's nothing more frightening than the belief that we're destroying the planet.

In fact none of the things being said and done by climate-change activists seem to be very scientific. Their thesis is remarkably simplistic and their argument is often led by media personalities and journalists who may sound knowledgeable, but usually know little about science; passionate political pseudoscientists (like Al Gore and Bill Nye the Science Guy, who ironically isn't a scientist), and popular celebrity scientists (such as Brian Cox and Tim Flannery), supported by a compliant and very unscientific mainstream/social media.

Their language is often scientifically 'loose' or just plain incorrect. Carbon dioxide is generally referred to as carbon (perhaps deliberately trying to create an image of something black and dirty). It's often called a pollutant, which it's definitely not (it's a colourless, odourless, naturally occurring gas we all breathe out which helps plants grow). And weather events are often called climate events and blamed on global warming, with no evidence whatsoever.

Most importantly, none of the alarmist predictions made twenty years ago have actually turned out to be true – not even close. All the climate models have been wildly inaccurate and incapable of replicating reality. As many of the more experienced scientists point out, computer modelling is a highly flawed process. It's certainly not science.

If the prediction of a seven-metre rise in sea level in the near future (made more than ten years ago by Al Gore in *An Inconvenient Truth*) was likely to be true, then surely there would be some indication of that happening by now. But in Sydney Harbour, one of the most land-stable parts of the world, tide records indicate the sea level has only risen by about 25mm (one inch) in the past 100 years. In fact, the sea level in Sydney Harbour actually dropped by 5mm from 1990 to 2018. [Fort Denison readings, www.bom.gov.au]

Science has to match reality. And very few of the alarmist's claims appear to be scientific. Yet in a weirdly Orwellian doublespeak kind of way, climate change alarmists are always claiming to be supported by science, while those who question their theory are supposedly 'science deniers'. Alarmists seem to be desperately trying to avoid scientific scrutiny by incessantly declaring that their theory is scientific 'fact' which, as such, doesn't need to be discussed, and certainly can't be questioned. This has been repeated so often that the general public has come to accept this too as a fact. When in reality it appears to be the complete opposite.

So what's going on?

Why would any group of people want to deceive us into believing we have to stop burning fossil fuels? And why would many of us be so easily deceived?

4. POLITICS & IDEOLOGY

"Political ideology can corrupt the mind, and science."
E. O. Wilson

At this stage, I think it's worth repeating the quote I used at the beginning of this book from Dr. Norman Doidge, author of *The Brain That Changes Itself*:

"Ideologies are simple ideas, disguised as science or philosophy, that purport to explain the complexity of the world and offer remedies that will perfect it. Ideologues are people who pretend they know how to 'make the world a better place' before they've taken care of their own chaos within. The warrior identity that their ideology gives them covers over that chaos. Ideologies are substitutes for true knowledge, and ideologues are always dangerous when they come to power, because a simple-minded I-know-it-all approach is no match for the complexity of existence. Furthermore, when their social contraptions fail to fly, ideologues blame not themselves but all who see through the simplifications."

There are many things in that quote which ring true in regards to the climate change movement. But I needed to find out more, so to get an understanding

55

of what's been happening politically and sociologically throughout the western world, I read dozens of relevant books, journals, and websites [listed in the references], and watched hundreds of hours of YouTube presentations and lectures by social psychologists and modern philosophers such as Canadian clinical psychologist and professor of psychology Jordan Peterson, American social psychologist Professor Jonathan Haidt, Canadian-American philosopher Professor Stephen Hicks, philosopher and best-selling author Robert Greene, American feminist academic and social critic Professor Camille Paglia, and American biologist and evolutionary theorist Professor Eric Weinstein.

Yes, I know, I could have been binge-watching *Game of Thrones* like most normal people, but what I discovered turned out to be not only fascinating, but also a vital key towards understanding the climate change movement.

I would describe all of the philosopher/social scientists I listened to as well-reasoned, middle-of-the-road social analysists with a great understanding of human psychology, human nature, and human history. As academics, they all describe themselves as left-leaning Democrat voters. So it's interesting that some of them (in particular, psychologist Jordan

Peterson) are described by critics as ultra-right-wing neo conservatives. In their defence, they claim that in the current social media/political environment, anyone to the right of Karl Marx is derided as a conservative. Which is something to keep in mind. Anyway, to the best of my ability, the following is what I figured out:

It looks as though the climate change theory just happened to align with a whole raft of political, sociological, philosophical, and environmental movements all coming together at the same time.

Climate change is not only the perfect vehicle for an authoritarian political movement that wants to restructure society, transfer power from oppressors to the oppressed, and redistribute wealth, but also provides an increasing number of atheists and agnostics with something to believe in – those who have abandoned traditional religions, but perversely, have adopted a new movement that has all the hallmarks of a religious cult – with human morality and sin (the greed, gluttony, and sloth of 'deniers'), uncompromising belief, and of course the inevitable Armageddon.

Add to this a widely-held post-modernist belief that capitalism and its associated consumerism is raping the earth, and a tendency to believe one's inner feelings rather than reason, rational argument, or logic – resulting in a scientifically naïve public who have been deceived into believing that CO_2 and air pollution are the same thing, and who really want to believe they can 'save the planet' by supporting a meaningful cause, which also happens to openly display their own moral virtue.

I know that there was a strong eco-movement against the burning of fossil fuels for many years before the CO_2 global warming theory became widely accepted in the 1990s. I had been part of it. It was based on two main concerns; depleting finite natural resources (especially after the seventies oil crisis), and air pollution.

The trouble was that the counter arguments to reducing fossil fuel consumption were frustrating environmentalist's demands for action. Opponents declared – quite rationally – that when oil and coal runs out, or it becomes too expensive to extract and process, another technology will gradually take over, and in the meantime we will have dramatically improved everyone's quality of life – people will be warm in winter, cool in summer, have fridges and washing machines and cars. And the rationalists would declare, *"So what if city air is a bit polluted … it's not the end of the world"*.

But the new theory of CO_2-driven global warming was a godsend for militant environmentalists. At last they could reply; *"well, actually, it will be the end of the world"*. Guilt and fear. Powerful allies.

And with the development of powerful new computers in the past few decades, a new discipline worked its way into academia. Computer modelling was the new boy in town. It was hard not to be impressed. Computers could churn out pages and pages of complex-looking graphs and charts and predictions, with hidden data, and assumptions, and algorithms that only the programmer understood, and could easily manipulate to get whatever result they were aiming for.

A computer model depends so much on the quality of the information fed into it, and the assumptions which have to be made, that often they provide nothing better than a wildly inaccurate guess, to the nearest decimal point – encouraging a dangerous level of hubris amongst young scientists who lack the wisdom and experience to know any better.

I've had personal experience with computer modelling programs used to predict the energy use of new buildings. The additional problem with them is that the more complex and sophisticated the programs have become, the more credibility they appear to have – yet the more easily the results are manipulated (inadvertently or deliberately) and they become even further removed from reality. This could be just what's happening with climate change modelling.

So it appears as though we've got three decades of over-ambitious scientists + a religious vacuum + a globalisation movement + anti-capitalist sentiment + post-modernist disdain for reason + militant environmentalism + computer modelling hubris + modern mainstream/social media that thrives on sensationalism.

This could be the 'perfect storm' of political, ideological, philosophical, and environmental movements that has made many of us susceptible to easily believe a rather simplistic theory that does not stand up to rigorous scientific analysis. Which could explain why its proponents have been trying to stifle debate by falsely claiming 'the science is settled'.

It's a political philosophy that can easily sneak up on a democratic society of fundamentally decent people who are predominantly caring and compassionate and concerned about the environment. But can also appeal to those driven by envy and ambition; who may claim to care about the environment, the poor, and the powerless, but in reality, are just clambering after totalitarian power for themselves.

It starts with an increasing acceptance of groupthink and political correctness, then differing opinions are redefined as 'hate speech', and disagreement is met with character assassination, which shuts down discussion or debate of important and often complex issues, and ends up with some really bad, poorly thought-out ideas being widely accepted by otherwise intelligent people.

A common tactic of this modern authoritarian movement is to shut down any opposition to their ideologies by subtle changes to the language and removing nuance from every issue, so there is in fact, no debate – the type of thought control and suppression of free speech which, according to some public intellectuals, is leading us down a path to the dystopian hell of George Orwell's *1984*.

Expressing an opinion even slightly out of line with current politically-correct ideology will get you publicly shamed, demonised, ostracized, and probably dismissed from your job, with media commentators clambering over each other to declare their outrage and position themselves on the high moral ground.

According to most of the modern philosophers I read and listened to, this is exactly what has happened over the past few decades with the environmental movement, which has become not only a socialist political movement, but also the perfect vehicle for moral posturing, virtue signalling, and social conformity.

After twenty years of being told by virtually every decent reasonable-sounding politician, commentator, environmentalist, and celebrity-scientist, that climate change is real, it's caused by CO_2 from burning fossil fuels, it's going to cause catastrophic sea-level rises and weather events that will irreversibly devastate the planet, making it the greatest moral challenge of our generation, and 97% of climate scientists agree; it's really hard to change your mind.

But beware of the pervasive and destructive power of groupthink:

"If this conformity becomes too ingrained in you, you will lose the ability to reason on your own, your most prized possession as a human." [Robert Greene]

However, if you really, truly believe in the impending apocalypse of human-induced climate change, and aren't interested in changing your mind, there's no point reading any further. In fact you probably haven't even read this far anyway. Nothing will convince you. On the other hand, if you're open to learning something new and really challenging your beliefs; and have the courage to think for yourself and even change your mind, then read on.

I dare you.

5. WHAT SCIENTISTS ARE SAYING

"Science at its best is an open-minded method of inquiry, not a belief system."
Rupert Sheldrake

Over a period of six months, I watched, listened to, and read leading scientists from both sides of the climate change argument. The following pages describe, as objectively as I can, the things I learnt and observed, many of which I had never heard before, despite being what I considered fairly well informed about climate change.

And I learnt some disturbing things about modern science. It seems that scientists really can sometimes get things terribly wrong – then either unconsciously or unscrupulously fudge their research to doggedly defend their position (and career). They often produce graphs and statistics which prove their point by manipulating scales, time spans, cherry picking data, and even using different data sets.

As Ben Goldacre demonstrates in his best-selling book *Bad Science*, it's quite possible (and frighteningly common) for scientists to manipulate statistics, and even supposedly legitimate peer-reviewed scientific research, into telling any story one likes.

"The psychological profile of these people is interesting," says Mario Biagioli, a professor of the history of science at Harvard University. *"You usually get B-plus, A-minus scientists who get into hyper-production mode."* In other words, we can end up with very ambitious, but very intellectually average scientists dominating certain fields. And sometimes 'science' is not very 'scientific' at all. It's susceptible to the same frailties, prejudices, and dishonesties that corrupt every human endeavour.

As most of us are aware, the general consensus appears to be that burning fossil fuels is increasing atmospheric CO_2, which as a greenhouse gas, is causing significant global warming and associated sea-level rises which will be catastrophic and irreversible in the near future.

But many experienced scientists claim that, firstly, there is no definitive scientific evidence of abnormal global warming or significant sea-level rises occurring in the past sixty years. They point out that computer modelling predictions do not constitute scientific evidence and are notoriously inaccurate.

Secondly, there's no evidence that the ongoing but gradual global warming that *is* occurring is caused by CO_2 emissions from burning fossil fuels.

And thirdly, it's just not that simple.

The earth's climate is a highly complex, dynamic system we barely understand. There are hundreds, if not thousands, of known and unknown factors affecting the earth's climate and atmospheric CO_2 levels (including changes in the earth's orbit, changes in the orientation of the earth's axis of rotation, plate tectonics and volcanic eruptions, greenhouse gases other than CO_2, ocean currents, vegetation coverage on the land, and of course the constantly changing solar output).

Alarmists claim that historical records show a close causal correlation between CO_2 and global temperatures. Other scientists claim that on closer analysis, the correlation is not consistent, and if anything, CO_2 follows a rise in temperature, not the other way around, as alarmists claim. So rising atmospheric CO_2 might be a response to rising global temperatures, not the cause.

Most scientists seem to agree that the earth has been warming at a rate of about 1°C per century since the Little Ice Age ended around 1850, with sea levels rising, on average, by about 12cm (5 inches) per century. Many scientists claim that in terms of the earth's changing climate throughout history, this is not unusual, unprecedented, or unmanageable.

According to many scientists, while the rate of warming and sea-level rise has not increased significantly in the last sixty years, during the same period, CO_2 emissions from burning fossil fuels have increased sixfold (from 5.98 billion tonnes in 1950 to 37.10 billion tonnes in 2018), which indicates the two factors may not be linked.

More specifically, twenty years ago when alarmists started warning us that if we didn't drastically reduce our consumption of fossil fuels sea levels could rise by up to several metres within a couple of decades, global CO_2 emissions were about 25 billion tonnes per annum. Since then, not only have we *not* reduced our CO_2 emissions, we've actually increased them by almost 50% to over 37 billion tonnes per annum. Yet during the same period, the average sea level readings at Sydney Harbour's Fort Denison actually dropped by 6mm (from 1998 to 2017).

Some scientists believe that global temperature records are simply not that accurate or reliable anyway, for many reasons. They are a mixture of tree-ring measurements, satellite readings since

1978, weather balloons since the 1950s, and surface thermometers in various parts of the world since 1850, many of which have been moved or are now surrounded by urban development. In fact some scientists claim that there is so much regional and local variance that there's no such thing as a 'global temperature'.

Alarmists claim that the ten warmest years on record have all occurred since 1998, which they say is proof that human-induced CO_2 emissions are driving up global temperatures. But other scientists point out that measured historical records only cover the last 100 years or so in most parts of the world, and records have only been scientifically reliable for the last forty years, since satellite readings began in 1978. But to the best of our knowledge, the earth has been in a warming trend ever since the Little Ice Age ended in about 1850, which of course means, quite logically, that every subsequent decade will be the warmest on record, as long as that trend continues.

There also seems to be clear historical evidence that the earth was warmer than it is now during the well-documented Medieval Warm Period (950-1100) when Vikings were growing barley in Greenland at a time when human-induced CO_2 emissions were insignificant. It also must have been significantly warmer when the Romans occupied Britain and were growing vineyards in northern England.

According to satellite data, there was an apparent pause in global warming from about 1998 to 2013, which incidentally was when alarmists changed their definition from 'global warming' to 'climate change'. Alarmists were quick to point out (quite rightly) that this does not indicate that global warming has ended, as it takes more than a fifteen-year pause to change an ongoing trend.

However, the pause was very significant for another reason. Because during the same period atmospheric CO_2 kept increasing at the same rate as usual (since we began measuring it in 1958), from about 370ppm in 1998 to 400ppm by 2013, while CO_2 emissions from burning fossil fuels increased significantly from about 24 billion tonnes/year in 1998 to 36 billion tonnes/year by 2013.

Which means that during a fifteen-year period when there was a 50% increase in CO_2 emissions, there was little or no increase in global temperatures. This could be seen as pretty solid evidence that something other than CO_2 is the main driver of global warming.

Atmospheric CO_2 has only been continuously and accurately measured (at Mauna Loa Observatory Hawaii) since 1958, and since then it has risen steadily and linearly from about 320 to 410 parts per million.

Estimates of CO_2 before 1958 come from Antarctic ice core samples which indicate that CO_2 concentrations dropped from 4,000 parts per million during the Cambrian period about 500 million years ago to as low as 180 parts per million during the Quaternary glaciation of the last two million years.

Alarmists claim that pointing this out is meaningless and misleading, since (obviously) humans didn't inhabit the earth 500 million years ago. But it's worth mentioning simply to indicate the scale of CO_2 variability during the earth's history.

It's also worth noting that commercial greenhouse operators usually aim to maintain a CO_2 level of around 2,000 parts per million to optimise plant growth.

Some scientists claim that historical estimates of CO_2 levels are very inaccurate anyway. They explain that tacking sixty years of actual atmospheric CO_2 readings onto a graph of estimates from ice core samples is misleading and scientifically problematic.

SOME UNDISPUTED FACTS ABOUT CO_2

Carbon dioxide (CO_2) is an odourless, naturally occurring gas that makes up about 0.04% of the earth's atmosphere. It is a minor greenhouse gas (a gas that contributes to the greenhouse effect by absorbing infrared radiation). The most significant greenhouse gas by far is water vapour (about 90% by volume). Other greenhouse gases include methane, nitrous oxide, ozone, chlorofluorocarbons, and hydrofluorocarbons.

Natural sources of atmospheric CO_2 include volcanic outgassing (on land and on ocean floors), the combustion of organic matter, wildfires, and the respiration process of humans and animals. The average adult human produces about 1 kg of CO_2 per day. Man-made sources of CO_2 include the burning of fossil fuels for heating, power generation, and transport, as well as some industrial processes such as cement making.

Plants convert CO_2 to carbohydrates by a process called photosynthesis. CO_2 is actually deliberately generated and pumped into commercial greenhouses to increase plant growth. So the recent rise in atmospheric CO_2 has resulted in significant global greening over the past century.

THE IPCC

"The Intergovernmental Panel on Climate Change (IPCC) is an intergovernmental body of the United Nations, dedicated to providing the world with an objective, scientific view of climate change and its political and economic impacts. It was established in 1988 with the task of assessing the risk of human-induced climate change, its potential impacts, and possible options for prevention". [Wikipedia]

Many scientists claim that the problem started right there, with the IPCC's stated objective; with the assumption that there is, as a matter of fact, *"human-induced climate change"* which has *"political and economic impacts"* and requires action.

And although it's hard to believe that an international government-backed organisation could be fraudulently driving a false agenda, this assumption, from a scientific point of view, is the fundamental reason it's all gone so terribly wrong.

The following is from an open letter in 2009 by Dr John Happs, a former lecturer in the geosciences and author of numerous science texts to Australia's then Chief Scientist Professor Penny Sackett:

"Some politicians still see the IPCC as being the gold standard of climate science. In fact the IPCC is a single-interest organisation that was established twenty years ago. Right from the start it assumed a widespread human influence on climate. Its charter was to assess the scientific, technical and socio-economic information relevant for the understanding of the risk of human-induced climate change. Such a charter makes it unlikely that the other factors influencing climate change would be taken seriously. In short, the IPCC's agenda appears to be political and ideological rather than scientific."
[www.undeceivingourselves.org/l-ipcc.htm]

The leaked (Climategate) emails in 2009 appear to reveal just how badly corrupted the IPCC has become.

Alarmists dismiss the email scandal by insisting they were taken out of context and reveal no more than *"simply a candid discussion of scientists working through issues"*. The university (East Anglia) investigated itself and *"concluded that we find that their rigour and honesty as scientists are not in doubt."*

The university was unlikely to vote itself out of existence by cutting off its major source of funding.

More from Dr Happs:

"Collectively the leaked material reveals serious abuse of the scientific process. The emails seem to reveal a clique of authors working covertly to ensure that only those papers supporting man-made global warming were published. In effect a small clique of scientists controlled the IPCC, the IPCC's crucial report chapters, and the IPCC's Summary for Policy Makers which went out to politicians and the media. The emails also appear to indicate that the clique manipulated data to favour the notion of unprecedented man-made global warming. Such manipulation of data, for political and/or ideological reasons, is misconduct at best and fraud at worst."
[www.undeceivingourselves.org/I-ipcc.htm]

The data the scientists used to come up with their conclusions about global warming has never been released (which apparently is most unusual in scientific circles), despite attempts through freedom of information requests and several court cases.

THE HOCKEY STICK GRAPH

The (now infamous) hockey stick graph, which purported to show a sudden and unprecedented 20th century warming, was an integral part of Al Gore's argument in his 2006 documentary *An*

Inconvenient Truth. It was also included in early IPCC reports but was dropped from later ones after its methodology and scientific integrity was questioned by many scientists.

More again from Dr Happs:

"Michael Mann's hockey stick graph appears to show that the Earth's temperature was stable from 1400 to 1900. There is then a dramatic rise (like the end of a hockey stick) that was claimed to be due to carbon dioxide emissions. It is now known that the data had been carefully fudged to remove an inconvenient truth, namely the Medieval Warm Period, when the world was warmer than today. Eventually the IPCC quietly dropped the hockey stick graph, claiming (contrary to the evidence) that the medieval warming was local and not global."
[www.undeceivingourselves.org/I-ipcc.htm]

In 2011 Michael Mann sued fellow scientist Dr Tim Ball for libel, after Ball had accused him of deliberate fraud in regards to the hockey stick graph. In 2019 a Canadian court dismissed the case and awarded all costs against Mann, who had refused to produce the data to prove the validity of his graph.

All Mann had to do to win his case was to produce the data.

The significance of this court ruling is huge but has gone mostly unreported by mainstream media. The hockey stick graph is the cornerstone of the climate change movement – apparent proof of a dramatic increase in global temperature in the past few decades – yet must now be considered highly suspect.

Another scientist, Berkley Professor Richard Muller, demonstrates in a brief and precise YouTube presentation that even without taking into account the Medieval Warm Period (which can't really be definitively proven one way or the other since there are no actual temperature records of that period), the scientists who produced the hockey stick graph fraudulently manipulated the data from 1961 onwards to create an apparent sudden rise in temperature.

EXTREME WEATHER

Despite the recent proclivity of alarmists and mainstream media to blame every hot day, cold day, storm, flood, drought, and bushfire on climate change, even the IPCC admits in its initial reports that there is no definitive evidence that there has been a worldwide increase in the frequency or intensity of these events in the past hundred years. Data released by insurance companies showing a steady increase in damage costs is misleading for obvious reasons; more buildings are continually being built, and costs are always rising.

Dr John Happs on extreme weather:

"As far back as 1996 the IPCC 'Science of Climate Change' report stated that 'it is not possible to say whether the frequency, area of occurrence, time of occurrence, mean intensity or maximum intensity of tropical cyclones will change.'

Since then the IPCC, no doubt prompted by advocacy groups, has become more alarmist about its extreme weather predictions, but without any supporting evidence.

In 2006 Wu et al found no increase in either intensity or number of hurricanes striking the USA and a significant downward trend for some areas of the Pacific.

2009 data from the Center for Ocean-Atmospheric Prediction Studies in Florida show that global tropical cyclone activity is currently at its lowest level in 30 years." [www.undeceivingourselves.org/I-ipcc.htm]

THE SCIENTISTS

Scientists on both sides of the climate change debate obviously believe in what they are saying. They have graphs and diagrams and data supporting their beliefs and opinions and seem, at least initially, equally convincing. So how does a layman decide who to believe?

We have no other option than to listen carefully with an open mind and try our best to figure out who has the most logical argument and make our own judgement based on the credibility, qualifications, experience, integrity, and motivation of each proponent.

Following are my observations:

Despite what their critics often claim, all of the so-called 'deniers' I have listened to and referenced in this book are qualified scientists who appear sensible and well-reasoned. Most of them hold or did hold senior positions in major universities or government bodies such as NASA. Some are even Nobel Prize winners, and many have actually served on the IPCC board. In general, they are older, often near the end of a distinguished career, and tend to make statements including the phrases 'it's very complex', and 'we don't know for sure'.

Being older and more experienced than most of their opponents, they perhaps have a better understanding of what science actually is, or at least is meant to be. This was clearly demonstrated when Dr John Christy was repeatedly interrupted by a young senator during his presentation in a US senate enquiry with the accusation:

"So you don't believe in climate change?"

Dr Christy replied testily:

"Senator, it's irrelevant what I believe in, as a scientist, it's my job to report what we've measured and observed."

The scientists labelled as 'skeptics' or 'deniers' don't seem to have a lot to gain from taking the stance they do. Many of them have been bullied, vilified, and in some cases forced out of their academic positions on spurious charges for daring to question the general consensus. Their opinions are not popular or politically correct, so none of them have garnered fame, fortune, or public adulation. Tim Ball apparently spent his life's savings defending himself against the libel charges which Michael Mann's legal team had managed to stretch out for eight years.

Alarmists, on the other hand, have had a completely different experience. They are lauded as moral crusaders, as saviours of our planet. Many have achieved fame, fortune, and lucrative careers. In general, they tend to be a generation younger than the 'skeptics', very certain they are right, and quite dismissive of their detractors.

And alarmist scientists always spend far more time talking about the consequences of catastrophic global warming than whether or not warming is actually caused by burning fossil fuels, which they take as a given. Usually their 'proof' relies solely on the repeated mantra that they are backed by scientific consensus ('97% of scientists agree', 'the science is settled') and their computer modelling projections …and of course, Michael Mann's infamous hockey stick graph.

Surprisingly, especially since alarmists are always claiming to be the ones with science on their side, the most outspoken alarmists, like David Attenborough, Brian Cox, and Tim Flannery are principally celebrity scientists from other fields, not working climate scientists.

In fact the most high profile alarmists such as Al Gore and Bill Nye are not even scientists at all.

Like many people who have believed the alarmist mantra for the past decade, I was convinced after watching Al Gore's 2006 documentary *An Inconvenient Truth*. It was an interesting exercise watching it again recently with a more critical eye.

Early in his presentation, after a few simple graphics which a six-year-old could easily understand, Gore announces: *"The basic science of global warming – I'm not going to spend a lot of time on this because you know it well."* Really? I certainly didn't. He then plays a segment from *The Simpsons* TV cartoon, which supposedly explains how we're meant to *feel* about global warming ... according to Lisa Simpson.

The rest of the documentary is dominated by apocalyptical scenes of hurricanes, floods, droughts, refugees, terrorists, and some lovely graphics showing Manhattan and Shanghai disappearing under a twenty-metre sea-level rise; with a lot of sentimentalism and moralising thrown in for good measure. You believe the world as we know it is coming to an end because you love your children. It's all good sensationalist stuff, and it convinced millions of people around the world – including myself – that we should be alarmed.

But it's not science.

6. THE REAL WORLD

"I have noticed that the most flexible, dynamic, inquisitive minds among my students have been industrial design majors. Industrial designers are bracingly free of ideology and cant.
The industrial designer is trained to be a clear-eyed observer of the commercial world;
which, like it or not, is modern reality."
Camille Paglia

Most of the information in this chapter comes from the research I did for my 2013 Masters dissertation, plus over twenty years of personal experience working with off-grid power systems. My dissertation was based on the following question:

What effect do grid-connected solar panels and wind turbines have on CO_2 emissions?

And after six months of fulltime research I concluded that not only is there no evidence to indicate that grid-connected wind turbines and solar panels actually lower 'real-world' CO_2 emissions, but with an understanding of the way an electricity grid works, there is no logical reason why they would.

Perhaps the most surprising thing is not even the conclusion I came to, but the fact that after searching thousands of books, articles, and scientific journals, I could not find one single reference to that question ever having been asked before. It apparently has just been assumed that connecting solar panels and wind turbines to a grid would automatically reduce global CO_2 emissions.

This is an astounding discovery. How could it possibly be so? And why has no one ever examined this before?

To explain it simply; because of the intermittent nature of solar and wind inputs, large fossil-fuel-powered generators must remain running in the background, no matter how many solar panels and wind turbines and batteries are connected to the grid. Some small gas-powered generators can be turned on and off at short notice, and pumped hydro and batteries can help manage the difficulties of accommodating wildly varying inputs, but with current technology, an industrial-size grid cannot operate reliably without the large majority of its energy coming from fossil fuels (or nuclear).

If you find that difficult to accept because of what you've been led to believe by authorities for the past twenty years, read the remainder of this chapter for conclusive evidence. I'm sure you'll be convinced.

Five years after completing my dissertation, it's become clear that the conclusion I reached was correct; connecting intermittent renewables to a grid achieves nothing other than a less reliable and more expensive electricity supply. South Australia and Denmark are perfect examples. They share the honours of the world's highest level of renewables, and the world's most expensive electricity prices. Only a politician or green ideologue could fail to see such an obvious connection.

THE DETAILS OF WHAT I DISCOVERED

To begin my dissertation, I first had to examine another question. Is it really possible that over the past few decades, trillions of dollars' worth of public funds might have been spent on developing, promoting, manufacturing, installing, and subsidising a type of power system that could never possibly work, on any level; a system that not only significantly increases electricity costs, but is also unreliable and difficult to manage and doesn't even lower CO_2 emissions anyway?

To believe this you'd have to be open to the possibility that not only can governments and bureaucrats get things completely wrong, and that corporations would act for no other reason than their own self-interest and short-term profit; but also, that it's possible for a large majority of the public to be totally ill-informed and misled by idealism, wishful thinking, and political movements with ulterior motives.

Recent events might sway many into believing this not only could happen, but most definitely is happening, in many different spheres of the modern world – such as the invasion of Iraq to prevent it from deploying its (supposed) secret cache of weapons of mass destruction; the way an unstable and unsustainable global financial system was allowed to develop, resulting in the collapse of several major banks which were given 'triple-A' credit-ratings by authorities only days before their demise [Preston, 2012]; and a modern food industry that has been allowed to chemically engineer products not to make us healthy, but to make us hungry and eat more, at the expense of our health – while telling us (with billions of dollars' worth of marketing) that it's good for us [Gillespie, 2012].

So the whole renewable energy industry could be based on a false premise – the premise that not only is it actually reducing CO_2 emissions, but if implemented on a large enough scale, it will be able to reduce emissions significantly enough to prevent global warming. It's quite possible that the industry could have been allowed to develop because governments make decisions based on their ability to attract votes from a misinformed public who desperately want to believe that grid-connected wind turbines and solar panels will 'save the planet', encouraged by an amoral corporate sector which is quite happy to go along for the ride, profiting all the way – regardless of its actual benefits to the environment.

Although most of us think we can tell opinion from fact, under a barrage of information, combined with an ever-increasing tendency towards political and corporate 'spin', it's getting harder and harder to tell the difference. The deeper one looks, the more one tends to find that many things presented as 'facts' aren't really facts at all – they're opinions, agendas, prejudices, and predictions, dressed up as statistics, scientific evidence, and academic reports [Goldacre, 2009]. This is the sort of information that forms people's beliefs, and sets public opinion, corporate policy, and political agenda.

When it comes to the environmental challenges the world currently faces, it can be difficult finding actual facts. For virtually every claim, it's possible to find research, statistics, and reports (by apparently equally well qualified authors) that indicate just the opposite.

A common distraction is to present facts that, while undoubtedly true, are irrelevant to the argument under discussion. For example, the fact that wind and sun are free natural resources does little to prove that they can help reduce an electricity grid's CO_2 emissions, or that they can actually reduce the cost of an electricity generating system.

The fact is, most facts are distorted and sifted through the filter of the interpreter's background, experience, and personal agenda. Which of course applies equally to this book and is why my background is explained in the preface.

Another problem is the (apparently widening) gap between scientists/academics/industry-specialists and reality. The sophistication, complexity, and apparent accuracy of computer modelling programs can be an alluring trap for those without regular contact with the 'real world' – to what is actually happening. To the extent that a group of modern architects and engineers can convince themselves they've designed a highly energy-efficient (and award-winning) building – because their computer modelling tells them so; despite the fact that when tested after completion (something the architects often try to avoid), the actual energy it uses is even higher than a normal building. [Lstiburek, 2008]. More often than not, fundamentals and common sense lead to a better result.

Environmental scientist James Lovelock eludes to yet other issues with modern science:

"Younger scientists cannot freely express their opinions without risking their ability to apply for grants or publish papers. Much worse than this, few of them can now follow that strange and serendipitous path that leads to deep discovery. They are not constrained by political or theological tyrannies, but by the ever-clinging hands of the jobsworths that form the vast tribe of the qualified but hampering middle management and the safety officials that surround them." [Lovelock, 2008. p.119].

Lovelock seems to be suggesting that a tendency of modern scientists to lean towards political correctness – for fear of damaging their career prospects by challenging the status quo – could be holding back genuine scientific enquiry. Additionally, just about everyone and every organization has an agenda – to sell a service or product, or simply to defend a belief system that many have invested a lifetime (or at least their career, and perhaps their self-identity) into developing. And ideological blindness – one of the greatest impediments to real progress – is rampant in the environmental, sustainability 'community'. Of course the question could be asked of me: What is your agenda? And I would claim it's simply to challenge my own beliefs and find out the truth. To support my claim, I would say that most of my life I have been an ardent supporter of any type of renewable energy source, grid-connected or otherwise. With additional research and more experience, I have changed my mind. And of course reserve the right to change it again if something, or someone, proves me wrong.

With all this uncertainty and conflicting information, perhaps it's even more important than ever to use logic and common sense (preferably backed up by as much relevant practical experience as possible) – combined with careful, open-minded observation – to help uncover the truth; or at least get as close to it as possible. And it's not a good idea to allow ourselves to be convinced by those who are too certain of their opinions. It's a complex, ever-changing world, and we only see a tiny part of it from our own perspective and background. In fact recent research has indicated that the more certain we are in our opinions, the more likely we are to be wrong.

In what philosopher Nassim Nicholas Taleb refers to as 'black swan events', we're often like turkeys – each day the farmer comes out to feed them they become more and more certain that he cares only for their wellbeing; until the fateful day he arrives with an axe. As James Lovelock says:

"A good scientist knows that nothing is certain; everything is a matter of probability".

Many of the conclusions I come to may be particularly challenging to those who already *know* what is environmentally/morally good and what is bad – the ones who 'believe' in climate change, the evilness of the fossil-fuel industry, and the virtues of wind farms and solar panels.

The trouble is, when people start *believing*, they usually stop looking and thinking, even when circumstances change and new evidence is unearthed. And what people believe may have little to do with truth or reality.

The fact that a few centuries ago most people believed the earth is flat didn't make it any less round. It just made it hard to open people's minds to what was actually quite obvious if they looked closely at the evidence.

According to former British Prime Minister Robert Peel:

"Public opinion is a compound of folly, weakness, prejudice, wrong feeling, right feeling, obstinacy, and newspaper paragraphs."

Which sounds just about right.

CO_2 EMISSIONS AND THE GRID

It's important to understand the effect of grid-connected wind turbines and photovoltaics on CO_2 emissions because it is usually the main focus of advocates for climate change action. The whole 'sustainability industry' is based on the premise that we urgently need to drastically reduce CO_2 emissions by generating all our electricity without using fossil fuels.

According to many people, wind and sun are the obvious answer. They are abundantly free and clean sources of energy that just need to be economically converted into electricity to save the Earth from human-induced climate change. What Americans would call a 'no brainer'; by definition, something requiring little or no thought. Which perhaps nicely sums up the current situation.

The following quote from an environmental website demonstrates many of the commonly held beliefs and attitudes towards renewable energy sources such as wind turbines:

"What's not to love? I'm an unabashed fan of wind turbines. I took a tour of a wind farm a few years ago and was blown away by the beauty of the towering, majestic monoliths spread out over the Wyoming hills. The slowly turning blades reminded me with every revolution that a little bit less coal would be burned to power our way of life. Wind power is environmentally friendly for numerous reasons. Emissions are negligible because no fuels are combusted, nor do turbines produce any substantial amount of solid waste while creating electricity."
[Mother Nature Network (www.mnn.com)]

In recent years, wind turbines have been favoured for large installations (predominantly rural and offshore wind farms) rather than photovoltaics because the bigger the turbines are, the more efficient they tend to be (as wind tends to be stronger and steadier higher off the ground), and the land under and around onshore wind farms can still be used for farming. Small wind turbines in urban environments tend to be ineffective (due to wind turbulence and interference from surrounding buildings), are sometimes noisy, and often unpopular with neighbours and local planning authorities [Boxwell, 2013].

The reality is that wind farms are the preferred option of power companies in order to provide the percentage of 'renewable' energy they require to satisfy government legislation simply because they are considerably cheaper per kilowatt than photovoltaics [EPA, 2013].

Photovoltaics (solar panels) tend to be favoured for smaller private installations because they can be made in small panels which are easily mounted on rooftops. Plus they are silent, can be relatively unobtrusive, have no moving parts, and require little maintenance.

They are usually sold to customers on the premise/promise that it will save them money on future electricity bills. However, the 'payback period' (the time it takes to recover the capital cost of the panels and installation) is very dependent on the government legislated 'feed-in tariff' (the amount the power company has to pay to buy back the electricity from the customer's solar system) [Boxwell, 2013].

The underlying – but generally unstated – premise is that by installing grid-connected photovoltaic panels, people are also 'doing their bit to save the planet' from global warming; although some solar-system advocates and designers are surprisingly frank about the real environmental benefits of small-scale grid-connected systems:

"If you have grid-tie solar but sell most of your energy to the utility companies during the day in summer and then buy it back to consume in the evenings and in the winter, you are making little or no difference to the overall carbon footprint of your home. In effect, you are selling your electricity when there is a surplus and buying it back when there is a high demand and all the power stations are working at full load. Do not assume that because you have solar panels on the roof of your house, you are automatically helping the environment." [Boxwell, 2013, p.13].

Nevertheless, most people in the renewable energy industry, along with most environmentalists, appear to be positive and optimistic about the role played by grid-connected wind turbines and photovoltaics in reducing the grid's CO_2 emissions.

Al Gore, in his book, *The Future*, makes the following claims, which are mirrored by most supporters of renewable energy:

"More potentially usable energy is received by the Earth from sunlight each and every hour than would be needed for all of the world's total energy consumption in a full year.

The potential for wind energy also exceeds the world's total energy demand several times over... Globally, renewables will be the second-largest source of power generation by 2015.

Almost half of the world's additional electricity generation will come from PV by midway through the next decade.

In the summer of 2012, there were periods when Germany received more than half its electricity from renewable energy sources.

In 2010, for the first time in history, global investments in renewable energy exceeded those in fossil fuels ($187 billion, compared to $157 billion)." [Gore 2013].

The International Renewable Energy Agency (IRENA) makes similar claims:

"With over 100 gigawatts of renewable power generation capacity added in 2011 alone, renewables have gone mainstream and are being supported by a 'virtuous circle' of increasing deployment, fast learning rates and significant, often rapid, declines in costs." [IRENA, 2013b].

The renewable energy industry makes it abundantly clear that not only are grid-connected wind turbines and photovoltaics an evidently sensible option, but they are well on the way to successfully transforming the world's electricity grids to CO_2-free systems within the next few decades. In fact, apparently, some countries are halfway there already.

This is undoubtedly heart-warming news for anyone who cares about the environment and is concerned about human-induced climate change. But amongst all the wonderful rhetoric, there is no attempt to address the issue of whether or not grid-connected wind turbines and photovoltaics are actually reducing CO_2 emissions.

The IRENA website contains hundreds of pages of information and over fifty published papers and journals; none of which make any attempt to examine or explain how or why grid-connected wind turbines and photovoltaics would reduce CO_2 emissions.

There is an obvious assumption that the more wind turbines and photovoltaics that are installed, the less CO_2 is being emitted. But no one is attempting to measure it. It's as though the original reason for doing something has been forgotten in the excitement of the implementation.

It's probably fair to say that most environmentalists (and much of the public) share the belief that unlike the people in the fossil fuel industry – who are only interested in profit and growth – the renewable-energy industry is owned and operated by people who care, and are only interested in the wellbeing of our children and our planet; helping to facilitate a transition to a 'sustainable energy future'. So of course they can be trusted to tell us the truth.

In reality, despite the veneer of social responsibility, the renewable-energy industry is a multi-billion-dollar industry, and undoubtedly subject to the same pressures to maximize growth and profit as any other big industry.

But environmentalists such as Al Gore obviously believe that anyone who isn't a supporter of renewable energy is an ill-informed conservative with a vested interest in the fossil fuel industry.

With this sort of prevailing image, it certainly wouldn't be very politically correct or career-sensible for a company CEO or politician to question the environmental credentials of the renewable-energy industry – an attitude which I suspect could be protecting the industry from proper scrutiny.

As for politicians, the *Annual Energy Statement* presented to the UK parliament in 2012 states that:

"The Climate Change Act 2008 established a legally binding target to reduce the UK's greenhouse gas emissions to at least 80% below 1990 base levels by 2050, and to achieve a 50% reduction in emissions over the 2023-27 period. ....renewables contribution increased to 3.8% of energy consumed in 2011 from 3.2% in 2010."   [DECC, 2012.p.10].

In the 43-page report, the UK government seems to accept the fact that electricity consumption will increase significantly (possibly double) in the next forty years, and that fossil fuel- and nuclear-powered generators will continue to provide the large majority (more than 95%) of the country's electricity. However, they plan to spend £12.7 billion on 'renewables' in the next year while making no attempt to explain how this will help them achieve an 80% reduction in emissions by 2050.

A large part of the report is dedicated to establishing the government's 'green' credentials and its commitment *"to prevent dangerous climate change"*, but there are so many contradictory statements in the report it's hard to take it seriously. The only thing that is clear about the UK government's energy strategy is that, after signing an international agreement to drastically reduce CO_2 emissions, they must at least *appear* to be taking action, no matter how convoluted, unexplained, and quite obviously impossible to achieve.

So while environmentalists and the renewable energy industry talk of a world rapidly approaching a situation where electricity grids are powered by close to 50% wind and solar, the UK government admits to a figure closer to 4%. It's as though they're living on different planets, with completely different realities.

Meanwhile, back in the real world, coal is by far the most widely used fuel for electricity generation on a worldwide scale, producing almost 39%, with gas next at around 23%, hydro at 16%, nuclear at 10%, and oil at 4%. Which means that almost 70% of the world's electricity supply is currently dependent on fossil fuel. Wind, solar, geothermal, and tidal combined provide a predominantly intermittent 8%. [IEA, 2018].

The following statements from an article by Dr Robert Peltier, editor-in-chief of *Coal Power,* an online industry magazine, give an indication of the current situation from the coal industry perspective, specifically in regards to Germany, one of the world's leading proponents of renewable energy:

"Germany is building more coal-fired power plants than at any time in the past 20 years for one very practical reason: They cost less to operate than the other options... The new coal plants were part of Germany's race to renewables ... There is one useful lesson to be learned from Germany's ruinous resource planning choices. Germany is building new ultrasupercritical coal plants designed to ramp up and down at 30MW/minute and 500MW within 15 minutes and shutting down older, less-efficient, and less-nimble plants. In other words, Germany's new coal fleet is designed to operate in a symbiotic relationship with renewables. The downside for Germany is that carbon emissions rose 1.6% last year as more coal was burned." [Peltier, 2013].

If nothing else, much of the information from Peltier is quite alarming to anyone who's spent years immersed in (and believing) green political rhetoric. He paints a very different picture to that depicted by environmentalists and the renewable-energy industry.

According to Peltier, the only thing that grid-connected photovoltaics and wind turbines have achieved is a massive increase in consumer electricity prices and an increased demand for coal-fired generators to back up the intermittent renewable energy sources. And the coal industry is quite happy to benefit from these consequences, regardless of the effect on CO_2 emissions.

Just like many of the claims made by environmentalists, the renewable energy industry, and politicians, the statements on a coal industry website could be selectively distorted to portray an optimistic future for the industry's shareholders. However, claims about the construction of coal-fired generators in Germany are easily verified from other sources with less obvious vested interests:

"Germany's dash for coal continues apace. Following on the opening of two new coal power stations in 2012, six more are due to open this year, with a combined capacity of 5800 megawatts, enough to provide 7% of Germany's electricity needs. Including the plants coming on stream this year, there are 12 coal fired stations due to open by 2020. Along with the two opened last year in Neurath and Boxberg, they will be capable of supplying 19% of the country's power." [Homewood, 2013].

The obvious question is: If wind and solar were working so well in Germany, why would they be building so many new fossil-fuel-powered generators? If the answer was that they need to quickly replace their recently and soon-to-be decommissioned nuclear power plants; that excuse could be countered by noting that most of the new generators were commissioned well before the government's populist political decision (soon after the Fukushima disaster of 2011) to phase out all nuclear power, and that coal-fired generators take many more years to build than wind turbines and photovoltaics [Peltier, 2013].

So why couldn't they replace the nuclear generators with renewables? It could be simple economics. Coal just might be the cheapest option. Or it could be that renewables simply add no (or very little) useable power to the grid, destabilise supply, and don't reduce CO_2 emissions anyway. It's most probably a combination of all these factors.

The one significant difference with claims made by the coal industry in relation to other interest groups is that they are quoting statistics and opinions from people working within the power industry, whereas environmentalists' and politicians' (and many renewable-energy-industry advocates) claims often sound like vague statements, wishful thinking, and political spin from people on the fringe of the industry, who are perhaps not as well informed (or ultimately responsible for designing and managing a reliable electricity grid) as they themselves believe, or would like us to believe. With a modicum of knowledge about how the grid, wind turbines, and photovoltaic systems work, it's hard not to be a little sceptical when proponents of this technology insist on quoting maximum installed capacity rather than the electricity actually usefully deployed by the grid. And according to the following article in *www.forbes.com* there's good reason to be sceptical:

"While the rapid growth in China's wind power capacity looks impressive on paper, it is less so in reality. China's total electricity production capacity reached 792.4GW at the end of 2008; the 12GW of wind capacity accounted for about 1.5% of that. However, in terms of actual power production, wind turbines generated 13 million megawatt-hours of electricity last year, only about 0.4% of China's total energy supply. China's wind turbine installation boom kicked off in 2006 as a result of a law that required power companies with over 5GW of production capacity to build enough non-hydro renewable power sources to make up at least 3% of their installed capacity by 2010, and at least 8% by 2020. However, the regulations do not stipulate how much energy must actually be generated from renewable power sources." [Wai-yin Kwok, 2009].

In Europe, according to a report titled *Wind Energy – The Case of Denmark* by the *Center for Political Studies* (CEPOS, 2009), an independent Danish research institution, a similar situation exists in Denmark, which is the world's leading proponent of wind energy, producing around 90% of the world's offshore wind turbines.
[The Official Website of Denmark, 2013].

Denmark has a policy of building no new carbon-emitting power plants (in Denmark), and a firm commitment to increase their wind-generated capacity to 50% of the country's electricity demand. It is currently claimed to be over 20%. But that is not to say that wind power contributes 20% of the nation's electricity demand:

"The claim that Denmark derives about 20% of its electricity from wind overstates matters. Being highly intermittent, wind power has recently (2006) met as little as 5% of Denmark's annual electricity consumption with an average over the last five years of 9.7%."   [CEPOS, 2009, p.2].

It's no coincidence that the Danes also happen to pay the highest residential electricity rates in Europe. [CEPOS, 2009, p.2].

In the previous section the question of why Germany would be building more coal-fired power stations was asked. A similar question could be asked about Denmark when observing their wind-turbine installation statistics. A hardened skeptic could suspect that Denmark discovered a serious problem with their own wind-based power strategy (which only works at its current level because of links to neighbouring countries with high levels of pumped-hydro storage capacities) as far back as 2001.

Wind generating capacity in Denmark has not increased since 2004. Since then they have apparently focused their efforts on manufacturing and selling wind turbines to other countries.
[CEPOS, 2009, p.9].

Apart from pointing out the grossly exaggerated input of renewable-energy sources, skeptics also claim that since photovoltaics and wind turbines are intermittent, feed-in systems do virtually nothing to reduce CO_2 emissions because the fossil fuel generators have to remain operating to provide a consistent base load.

They claim that for every megawatt of renewable energy added to the grid, the equivalent is needed as backup, or what is known as 'spinning reserve'. A full capacity of 'other' electricity must be available at all times in case the wind stops blowing or the sky suddenly clouds over. So according to skeptics, the 'wind and solar revolution' has done nothing more than appease the green vote, waste enormous amounts of money, increase energy bills, and destabilise the electricity grid. And it may actually increase overall CO_2 emissions. [Rosenbloom, 2012]

In *Wind Report 2005*, E.ON Netz (the grid manager for about a third of Germany, hosting 7,050 megawatts of wind-generating capacity at the end of 2004) states:

"Traditional power stations with capacities equal to 90% of the installed wind power capacity [a little over the maximum historical wind power infeed] must be permanently online in order to guarantee power supply at all times. The consequence is that wind power construction must be accompanied by corresponding construction of new conventional power plants. The result is that, while wind-generated power itself is CO_2-free, the saving to the whole power system is not proportional to the amount of fossil-fuelled power that it displaces."

In a presentation to a conference in Copenhagen on May 27, 2004, Flemming Nissen, the head of Danish generating company *Elsam*, stated that:

"Increased development of wind turbines does not reduce Danish CO₂ emissions." [Rosenbloom, 2012]

Similarly, Richard S. Courtney explains in a presentation to the 2004 *Groups Opposed to Windfarms in the UK* conference that windfarms only force power stations to switch more often between generation and spinning reserve, or standby:

"They provide no useful electricity and make no reduction to emissions from power generation. Indeed, the windfarm is the cause of emissions from a power station operating spinning standby in support of the windfarm." [Rosenbloom, 2012]

Some academic research journal articles express a similar opinion:

"Frequent ramp ups and downs of coal-fired plants lead to lower energy efficiency and higher emissions." [Li, et al, 2012]

This is a concept that most people would understand. They know that their car engine operates more efficiently and achieves more miles per gallon at a steady speed on a highway than at variable speeds in the city.

Apparently grid generators are similar.

Germany is leading the world in its innovation and implementation of renewable energy, with a claimed input of 25% of total grid demand by 2012. But the following article from the Institute of Energy Research (2013) points out some of the consequences for Germany's electricity grid:

"The government's transition to these intermittent green energy technologies is causing havoc with its electric grid and that of its neighbours – countries that are now building switches to turn off their connection with Germany at their borders. The intermittent power is causing destabilization of the electric grids causing potential blackouts, weakening voltage and causing damage to industrial equipment. These power grid fluctuations in Germany are causing major damage to a number of industrial companies, who have responded by getting their own power generators and regulators to help minimize the risks." [Institute of Energy Research, 2013].

Smaller gas- and coal-fired generators, that can be turned on and off relatively quickly, are apparently being built to facilitate the integration of intermittent renewables. But according to the following report from Imre Gyuk, who manages the Energy Storage Research Program at the U.S. Department of Energy, they are not helping to reduce CO_2 emissions:

"Meeting wind integration requirements with fossil generation will result in added emissions associated with part-load operation of thermal plants when they are placed into the duty cycles needed to support renewables integration ... wind generation is expected to present a challenge to grid operators at increasing penetrations.

With the grid already scrambling, it's hard to imagine adding more renewables, like wind and solar power, because they are intermittent sources of power.

We know customers are unpredictable, but now, so is the electricity. When the wind dies unexpectedly, a wind farm can lose 1,000 megawatts in minutes and must then quickly buy and import electricity for its customers. The alternative then is to use a peaker-style fossil-fuel plant, but that adds air pollution to clean electricity.

These expensive fossil-fuel plants sit idle all year and can emit more air pollution than a large coal-fired plant. If the peaker plants fall short, utilities pay large customers like aluminium smelters to use less electricity. If nothing works, you have brownouts and rolling outages." [Nasr, 2009].

It sounds very much like South Australia, circa 2019.

110

A website by an organization called *The Energy Collective*, *"an independent, moderated forum of the world's best commentary & analysis on energy policy, climate change, energy technologies and fuels, and energy innovation"*, whose members *"are our content contributors, and include leading scientists, activists, policy makers, executives and entrepreneurs"* looks to be a reliable source.

An article posted by research scientist Schalk Cloete claims that renewable energy other than hydro will contribute only about 5% of the global energy mix by 2035. It reinforces the view that the input and potential of grid-connected photovoltaics and wind turbines is vastly overstated – painting a very different picture to that depicted by the renewable energy industry.

Some more extreme comments from wind-farm opponents make interesting reading:

"Enron invented the modern wind industry by buying the support of environmental groups for large-scale 'alternative' energy and all that makes it profitable: tax avoidance schemes, public grants and loan guarantees, artificial markets for 'green credits', and laws requiring its purchase. Wind energy is just one more extractive industry, and with the collaboration of Enron's environmentalists it opens up land

normally off limits to such development. Industrial wind development may not be the worst scourge on the planet, but that does not excuse it. It is, however, particularly evil because it presents itself as the opposite of what it is." [Rosenbloom, 2012]

Although it wouldn't be the first time in history that large corporations have manipulated government legislation and public opinion for their own benefit, the interpretation above may or may not be completely true. But such extreme rhetoric shouldn't be used as an excuse to dismiss all of the skeptics' arguments, many of which are logical and backed up with sound research and references to legitimate government and industry reports.

In fact the skeptics appear to be the only ones looking critically at the issue. All of the detailed information about how renewables integrate with the grid is provided by critics – it is not even mentioned by wind and solar advocates. On close examination, the 'feel-good' rhetoric from environmentalists and the renewable energy industry is mostly based on mistruths, exaggerations, and the dubious argument that because more and more money is being spent on renewables each year they must be lowering CO_2 emissions, which supposedly is the only reason to implement them.

In his 2013 book *The Future*, Al Gore boasts that:

"Globally, renewables will be the second-largest source of power generation by 2015".

By 2020 this claim has turned out to be so far from the truth it makes it hard to believe anything he says.

Opponents of renewable energy technology are often dismissed as either money-hungry fossil-fuel-loving conservatives or self-centred NIMBY (Not In My Backyard) activists who aren't prepared to make even the slightest personal sacrifice (in lifestyle, taxes, convenience, or aesthetics) for 'the common good'.

NIMBY activists generally focus on selfish, subjective, and sometimes exaggerated issues such as noise, land values, aesthetics, personal health, and birdlife. But it should also be pointed out that most of the academic research journals reviewed for this book that purportedly analyse the pros and cons of wind generation focus only on these issues too. They make no mention of the broader (and far more important) issues of grid-integration, storage, and overall CO_2 emissions.

As windfarm advocates correctly point out, large, slow-moving turbines can look beautiful, depending on your attitude. And there are probably infinitely more birds killed each year flying into windows and hit by vehicles than are ever likely to be killed by the blades of wind turbines, no matter how many are built. But these issues have little to do with the main problem. More to the point is that the skeptics (the ones who aren't simply NIMBYs) are the only group publicly asking the critical questions.

And unlike the renewable energy industry, politicians, the fossil fuel industry, and environmentalists, these skeptics appear to be the only ones without a clear agenda of self-interest or political dogma. They occasionally sound frustrated and angry, but in general appear to have more integrity than other groups, and are at least thinking and observing; not dreaming, blindly believing, and manipulating public opinion through fear and ridicule with emotive arguments.

GRID STORAGE

As pointed out earlier, the main reason that wind turbines and solar panels may not be actually reducing CO_2 emissions is that backup generators have to remain running to provide a reliable baseload. The electricity wind and solar generate is unreliable, intermittent, and stochastic (randomly variable). The wind might not blow for weeks on end; the sun might shine brightly one minute, but not the next. So the electricity they generate can't be supplied 'on demand', when it is needed. And when it isn't needed, and can't be stored, the electricity is wasted.

Additionally, sudden variations (when a cloud passes across the sun for instance with photovoltaics) can cause havoc with the grid voltage, which makes it difficult to manage and can damage electrical equipment. Without storage, conventional generators have to remain running in the background ('spinning reserve') to maintain a steady and even supply of electricity.

If this is indeed the case, then the only thing achieved by connecting wind turbines and photovoltaics to the grid is a vastly more complex, unstable, and unreliable electricity grid, at an enormous cost – with little or no reduction in CO_2 emissions anyway.

Although the power industry is well aware of this fundamental problem – that mass storage is required in order to integrate a large proportion of intermittent renewables into the grid – and is working towards finding solutions (discussed later in this chapter), it is rarely acknowledged or openly addressed by environmentalists or the retail side of the renewable energy industry. In fact, the current lack of genuine, large scale grid storage could be described as 'the elephant in the room' of the renewable energy industry.

So while 'grid parity' (when the cost of renewable-energy-generated electricity is reduced to that of conventional grid electricity) has been the Holy Grail of the renewable energy industry for many years, and supposedly the only thing really holding back a wholesale transformation of the world's electricity grids (Wikipedia: Grid parity, 2013), it may not be the real problem after all.

"Energy storage is the key enabling technology for renewables. Until you can make [energy storage] reliable and affordable, it doesn't matter how cheap and efficient you can make wind and solar, because our grid can't handle the intermittency of those renewable technologies". [Buie, 2013]

The uptake of household roof-mounted photovoltaic panels took off in many countries in recent years with the advent of the grid-connected system with a generous (usually heavily subsidised) feed-in tariff, which forced energy companies to buy back electricity from their customers at a government-legislated price usually higher or equal to the retail selling price.

Although not popular with electricity retailers (it's obviously not sustainable from a business perspective – no business wants to be forced to buy its own produce off its own customers), this not only made it a financially viable option for householders, who theoretically would receive free electricity for the next twenty years and could payback the installation cost within a few years, but it apparently overcame the problem of intermittent-renewable-energy storage; because households could draw off the grid at night when their solar panels weren't producing any electricity [Boxwell, 2013]. This is often referred to as 'grid storage'.

But in reality, 'grid storage' is a rather deceptive play on words when used in this context. The grid is just a system of wires, poles, and transformers. The best it can do is shuffle electricity around to even out supply and demand (what is now being promoted as a 'smart grid'), but the grid itself, even a 'smart' one, cannot store or create electricity [Smart Grid, 2013].

A grid will never be able to absorb or 'store' a large amount of solar- or wind-generated electricity unless some sort of genuine storage is built into it on a massive scale. And this appears to be the limiting factor for grid-connected intermittent renewable energy.

"At present, renewable variability is handled almost exclusively by ramping conventional reserves up or down on the basis of forecasts. However, as renewable penetration grows, storage and transmission will likely become more cost effective and necessary. [However] Energy storage on a utility-scale basis is very uncommon and, except for pumped hydroelectric storage, is relegated to pilot projects or site-specific projects." [APS Physics, 2013].

Water pumped to a height (using solar- or wind-powered pumps) and stored in large reservoirs, then released through generating turbines when electricity is required, is a method of 'storing' renewable electricity. Pumped storage is the largest-capacity form of grid energy storage currently available, and, as of March 2012, the Electric Power Research Institute (EPRI) reports that Pumped Storage Hydropower (PSH) accounts for more than 99% of bulk storage capacity worldwide, representing around 127,000 megawatts.
[Wikipedia, 2013: Pumped-storage hydroelectricity]

In fact *"Pumped Storage Hydropower (PSH) is the only conventional, mature commercial grid-scale electricity storage option available today."*
(National Hydropower Association, et al, 2010)

Hydropower, which is a form of renewable energy in itself, provides around 16% of the world's electricity. But not all hydroelectric schemes have the ability to pump water back up to their reservoirs, which are filled naturally from rainfall in catchment areas. So although they are effectively storing energy, and can help even-out supply provided by wind turbines and photovoltaics by generating 'on demand' as part of a smart grid, the majority of hydroelectric schemes cannot actually store the electricity generated by wind turbines and photovoltaics.

As a result, most countries in the world have only a very small capacity, if any, for pumped storage – with limited potential for expansion.

Hydroelectric schemes are built at significant ecological expense [EPA, 2013a]. Dry countries like Australia often need the water downstream of large reservoirs or they risk destroying river ecosystems and irrigated farmland, and most developed countries have already fully exploited their river systems.

Australia, for instance, already has 502 major dams, has only built a few in the last couple of decades, and is unlikely to build any more.
[The Earthmover & Civil Contractor, 2010].

Although pumped hydro storage (PHS) will probably be increased by a few per cent each year in many countries over the next few years [National Hydropower Association, 2010], at this rate of expansion (which obviously cannot continue indefinitely) it is unlikely to ever provide anywhere near enough storage to enable the integration of a large percentage of photovoltaic- and wind turbine-generated electricity into the grid.

As for other storage options, a 2010 paper titled *Electricity Energy Storage Technology Options* by the *Electric Power Research Institute of California* [EPRI, 2010] concludes that compressed-air and advanced-battery technology are the two most practical and affordable options with the best potential to significantly increase grid storage.

But these are still in the development phase and have yet to be commercially implemented on any significant scale.

As with all electricity storage options, their feasibility seems improbable (in terms of money and resources) when the enormous scale of the storage required to make a grid anything close to emission free is realistically considered.

HOW A MICRO-GRID WORKS

Until about thirty years ago, before the advent of commercially available photovoltaics and small wind turbines, the most popular option for a continuous supply of electricity in a remote location was a continuously running diesel-powered generator.

And anyone who's had to live with this type of system knows that it doesn't matter how many lights or appliances you turn off, it makes very little, if any, difference to the amount of fuel being consumed (the system's CO_2 emissions). The generator is either running, and you have electricity, or it's not, and you don't. In fact it's usually best to have some appliances (or a dummy load) on continuously to keep the diesel engine under a steady load.

Even if no appliances are turned on, the generator must remain running so that electricity is instantly available when something is turned on. This is a fact apparently misunderstood by many people, including most advocates of grid-connected renewables. CO_2 is emitted when electricity is generated (by fossil fuels), not when it is consumed. When a generator is running, and electricity is available, the electricity may or may not be drawn off – it's just there, waiting to be used, it doesn't go anywhere if it's not used.

Keeping this in mind, it would be of no benefit to feed extra electricity into the system from solar panels or wind turbines, which essentially would be no different from turning off a few appliances. All it would achieve is to make the whole system difficult to manage, highly unstable, and possibly damaging to some appliances.

With wild fluctuations already possible on the demand side (perhaps a 1.2kW hairdryer, a 1.5kW iron, and a 2.4kW vacuum cleaner all suddenly turned on at exactly the same time), introducing wild fluctuations on the supply side for no actual benefit would be ridiculous – which is what would happen when, say, the 5kW from a photovoltaic system suddenly drops to 100 watts as a dark cloud passes across the sun.

For a small domestic power system (essentially a micro grid), if your maximum electricity demand (i.e. everything was switched on at once) was say, 10kW, you would probably install a 15kW diesel generator, which allows a bit up your sleeve to handle extra surges from some appliances which draw more power than their continuous rating for a few seconds when initially turned on (electric motors in particular), plus any random additional appliances someone might plug in.

In theory, you could instead install a 10kW diesel generator plus a 5kW wind turbine and photovoltaic array to handle the total 15kW demand, some of the time. But of course you would also have to install another 5kW diesel generator, and keep it running continuously, because the input from the renewable energy sources is going to be fluctuating from zero to 5kW on a minute by minute basis, and considerably less than 5kW most of the time (especially at night).

Theoretically, if you could accurately predict when the total demand was going to be less than 10kW at certain periods of the day or night, you could turn the 5kW generator off for a while. But it's a lot of complexity for a marginal saving in fuel. In fact the larger 15kW generator just running continuously, with a lower loading much of the time, would probably be more economical overall.

By adding the wind turbines, photovoltaics, and extra generator, you've just created a very complex, difficult to manage system – at enormous additional cost – for no apparent benefit, and little if any reduction in CO_2 emissions. In fact, if the embodied energy of building wind turbines and solar panels is taken into account, CO_2 emissions are probably significantly increased. And the more wind turbines and solar panels added, the worse it gets.

Of course a large municipal electricity grid is different to a simple micro grid, but it's worth noting that the problems identified above are exactly what seems to be happening to main grids which have tried to integrate wind turbines and photovoltaics.

UNDERSTANDING A LARGE GRID

So why does a move towards grid-connected renewable energy (wind and solar) actually increase electricity prices?

To explain it as simply as possible, if you owned a large coal- or gas-powered generator that supplied all the electricity to your local region, but the government imposed a 50% renewable energy target, which they reach by encouraging (with subsidies, financial assistance, and market-share guarantees) a renewable energy company to build a large solar panel/wind farm; it would mean that on windy sunny days the grid would only buy half of the power off you for much of the day. Unless of course the wind suddenly drops or it becomes overcast (and at night) - in which case you must have that power available for the grid if and when it's needed.

So your running costs (and CO_2 emissions) are pretty much the same as before any renewables were introduced to the grid, and you'd have to charge more for your power because they are buying less of it and there's no one else you can sell it to. And the more renewables they connect the more you have to put your prices up. But they tell us that electricity prices should drop because wind and solar are free. Which of course is not true in a practical sense.

If you've just built a large wind or solar installation, you've made a considerable capital outlay which you need to make a return on, you've got continual maintenance costs, and you probably have to pay rent on the land you're using. So you won't be giving away your 'renewable' electricity for free. You charge the grid for what they use, at market rates. The grid also has to pay for the electricity they get from consumers roof-top solar panels.

It's obvious that feeding electricity from solar panels and wind turbines into a grid doesn't make sense, at any level. It's not reducing CO_2 emissions, and it's increasing operating and infrastructure costs. In fact it's an idea being promoted by people who must be completely ignorant of technical reality, and/or whose only concern is to give the impression they are doing something to 'save the planet'.

Scientists do science, but engineers design generators and electricity grids. So when you hear a scientist, or even worse, a politician or green activist, declaring that a major electricity grid will be close to 100% emission free (using renewable energy) within the next twenty years, you can be sure they have no idea what they're talking about.

I know how difficult and expensive it is to power a building with renewable energy – even a relatively small, simple, but extremely energy-efficient home. My house utilised the latest solar and battery technology, but could only be made to operate on 100% renewable energy by reducing its daily energy consumption to less than a third of a normal house.

To achieve this, it had to be extremely thermally efficient, and it had to be a specific design which limited the amount and orientation of glazing and the house's overall internal volume to minimise the heating load in winter and heat gain in summer.

Even without taking into account the additional expense, this sort of energy efficiency simply can't be achieved with existing housing stock, or with most standard house designs. The house site also had to have clear solar access all day, with no shading from trees or surrounding buildings, especially throughout winter, which can only be achieved on a small percentage of building sites throughout the world.

So if it's this difficult and expensive to make one small, super-energy-efficient house totally emission free, what would it take to make a large interconnected national electricity grid emission free; powered by renewable energy with battery storage?

The first problem is that large electricity grids are grossly inefficient and incredibly leaky, with transmission inefficiencies possibly as high as 50%. The second problem is that the large majority of buildings are not energy efficient in any way, and it's prohibitively expensive, if not impossible, to make them so. And the third problem is that heavy industry uses electricity on a scale that dwarfs the capabilities of wind turbines, solar panels, and batteries, even on windy, sunny days. Current technology batteries cannot provide anywhere near enough storage to be of any practical significance to a large grid.

The world's largest latest-technology battery has just been built in South Australia by Tesla. But it can only supply less than 3% of the state's peak demand for about one hour. And of course it doesn't generate any electricity, it only stores it. Politicians and TV 'experts' often refer to it as a 129MW power system, which is very misleading. It's a 129MWh *storage* system. In other words it can provide 129MW for

127

one hour. A 129MW fossil-fuel-powered generator can provide 129MW continuously, every hour. That's 3,096MWh every day. So despite the media and political hype, the only real function of South Australia's very expensive new battery is to even out the variable supply fed into the grid by wind and solar.

To be 100% powered by renewables, a quick calculation indicates that Australia's national grid demand of 35,000 megawatts would require more than 5,000 battery banks equivalent to the 129MWh Tesla battery. At an estimated cost of $240 million each, the total cost would be $1,260 billion.

That's almost equivalent to Australia's total GDP, for batteries alone, without even considering the cost of solar panels and wind turbines capable of producing 35,000 megawatts continuously for at least eight hours a day. You'd also have to replace those batteries every 10 years.

Even if you reduced the battery costs to a fraction of their current price it still wouldn't be anywhere near viable. But that's not even half the problem, because you've then got to build another power system to recharge those batteries the next day (additional to the system that's powering the grid during the day), after they've been drained flat overnight. And what if the next day isn't sunny or windy?

In summary, it simply doesn't make sense to believe that by continuing to add solar panels and wind turbines and batteries to the grid, we will eventually have a reliable emission-free electricity supply. It's like force-feeding more hay to a packhorse in the belief that it will eventually be able to carry a thousand times its current maximum load.

MEASURING CO_2 EMISSIONS

Obviously if we are serious about lowering CO_2 emissions, we would want to measure them in some way in order to know if our actions are having the desired effect. There are three ways in which CO_2 emissions are commonly estimated or measured. They are listed below in increasing order of significance, accuracy, and connection to reality:

1. Indicative CO_2 Emission Estimates
Statements such as *"replacing one regular light bulb with a compact fluorescent light bulb will save 150 pounds [68kg] of carbon dioxide a year"* [UNESCO, 2013] are commonly made by environmentalists. These estimated reductions are just to give people a way of comparing how different activities and

appliances might affect CO_2 emissions (theoretically). But they can be very misleading. As pointed out earlier, CO_2 is emitted when electricity is generated by fossil fuel, not when it is used by the consumer. So turning off a light doesn't actually lower your CO_2 emissions. The electricity supplier is not going to turn a generator off because you've turned a few lights off. Nor are they going to turn a generator off if thousands of households turn off a few lights – because everyone might suddenly turn them on again and the electricity must be available instantly when they do.

2. Actual CO_2 Emission Estimates

These figures are also estimates, not actual emissions, so they are not necessarily real, or accurate, and can vary quite significantly depending on the parameters, methodology, and assumptions used.

An article by Zeke Hausefather on a website called *The Yale Forum on Climate Change & The Media* discusses CO_2 emission estimates. He claims that recent carbon reductions in the U.S. were brought about by a combination of factors, including; driving less, improved fuel efficiency, flying less and more efficiently, switching from coal to natural gas, overall energy efficiency, and wind generation. He makes no mention of photovoltaics. On the other hand, the United States Environmental Protection Agency [EPA, 2013b] states that United States CO_2 emissions increased by just over 8% from 1990 to 2011, although they did decrease by around 1% from 2011 to 2012.

They attribute this reduction mainly to the switch from coal to natural gas combined with an overall decrease in electricity demand resulting from a warmer than usual winter. There are several other reasons listed as possible influences on CO_2 emissions. *"Non-fossil alternatives"* gets a mention as a minor influence.

3. Actual Fossil Fuel Consumption

According to the Environmental Protection Agency, *"fossil fuel use is the primary source of CO_2"* [EPA, 2013b]. So even though it's not a direct measure of actual CO_2 emissions, a reduction of fossil fuel consumption might lead to, or indicate, a reduction of emissions. And if grid-connected wind turbines and photovoltaics were replacing fossil fuel-powered generators by any significant degree, then we could expect to see some reduction in fossil fuel consumption. According to the BP Statistical Review of World Energy (June 2012):

"World primary energy consumption grew by 2.5% in 2011, roughly in line with the 10-year average. Consumption in OECD countries fell by 0.8%, the third decline in the past four years. Non-OECD consumption grew by 5.3%, in line with the 10-year average." [BP, 2012].

It seems that the only time growth in world energy consumption has slowed in recent years was in 2008, almost certainly caused by a downturn in economic activity during the global financial crisis.

So despite trillions of dollars invested globally in the past two decades on grid-connected renewables, the two most authentic indicators of CO_2 emissions show no indication of the slightest reduction. A fact which is even more significant when all the other influences which might have helped reduce world CO_2 emissions in recent years are taken into account: including a major world financial crisis causing decreased production in many countries, a significant switch from coal and oil to natural gas, billions spent promoting and implementing energy efficiency in buildings, and improvements to vehicle efficiency and emissions.

Could it really be possible that a whole multi-billion-dollar industry has been built on little more than generous government subsidies and legislation from politicians keen to win votes by appearing 'green' to an idealistic, well-intentioned, but grossly deceived public?   An innocent population who believe that even our very existence is adding noxious global-warming-glacier-destroying-cyclone-producing carbon dioxide to the atmosphere; with every breath, every drive, every self-indulgent holiday; not only robbing our children of a future, but destroying a whole planet.

What else but such existential guilt and moral angst could possibly make otherwise intelligent people be so easily sold an idea that clearly doesn't work; simply because we so desperately want to believe that it does, that it will save the planet.

A recent article in *The Times* (UK) sums it up perfectly:

"Through age, belief and disposition, I'm the sort of guy who sees a wind turbine and feels a tingle in his spine. It's something to do with the glorious meeting of natural beauty and human ingenuity, audibly subdued and at the scale of a cathedral. There's a set I pass often, heading out of Edinburgh on the A68, dotted along the horizon of the Lammermuirs. They make me want to park the car and start singing hymns. And probably because of that I get quite

upset when people start arguing that they don't work terribly well. I can just about cope with an economic criticism; that we have the incentives wrong and wind farms are actually subsidy farms, rewarding technological box-ticking rather than the actual generation of power. But start advancing the notion that the whole concept is just not a very good idea – that, like hybrid cars or most solar panels, wind turbines are all for show – and I start feeling edgy. Never mind, for now, whether such claims are true. The important thing is how badly I want them not to be." [Rifkind, 2013, p.23].

Could all those politicians, scientists, engineers, and technocrats really get it that wrong? Could it be that for all their education, intelligence, and computer-modelling skills, they have no real practical knowledge of how electricity grids, generators, wind turbines, and photovoltaic systems actually work together in practice? Or are they simply suffering from an overwhelming case of ideological blindness?

At least one prominent scientist believes they've all got it terribly wrong:

"Enthusiasm for renewable energy coupled with a politics in which each nation tries to gain brownie points for its diligence in meeting the Kyoto limits is an unhappy mixture. It will fail and bring discredit

both to the greens and to the politicians. The responsibility for the wrong advice given to the government came from well-meaning city dwellers with a romantic, impractical dream of clean renewable energy. It is a consequence of the vulnerability of people to the astonishing power to deceive of an endlessly repeated falsehood." [Lovelock, 2007, p.107].

Estimates vary enormously about how much energy is currently and potentially provided by 'renewables'. The best realistic estimate I could find (from the EPA) is that currently, less than 2% of the world's energy is provided by wind turbines and solar panels.

Claims of 40 or 50% renewable energy in places like South Australia and Denmark are highly misleading. Firstly, because the electricity grid is only part of the overall energy demand (about 80% of the world's energy consumed is required for heating and transport, which are predominantly powered by fossil fuels).

Secondly because, as stated earlier, advocates usually inflate the importance of renewables by stating the maximum potential installed capacity, which is always way more than what is actually utilised. [A 100kWp solar array (the 'p' stands for peak) will only generate close to 100kW on a perfectly clear day, when the panels are near new and perfectly clean, and the sun is perpendicular to the panels. Most of the day it will only generate a tiny fraction of its peak capacity].

And thirdly, because they can't achieve such a high level of renewables penetration without a backup connection to a compliant fossil-fuel- or nuclear-powered neighbour which happens to have a lot of spare capacity when they need power on demand.

Outspoken scientists such as Lovelock (2007) and Flannery (2005) believe that we need to reduce the world's CO_2 emissions by at least 80% to prevent catastrophic global warming. But even if that were true, it's clear that connecting more and more wind turbines and solar panels to electricity grids will not do anything towards achieving such a lofty goal.

Fortunately, as I've recently discovered – according to better informed and less ideologically-driven scientists, and reality – it's not true anyway.

7. MORE HARM THAN GOOD

"Whenever a doctor cannot do good,
he must be kept from doing harm"
Hippocrates

There's strong evidence to suggest that some, if not all, of the so-called climate change action and rhetoric of a 'climate emergency' in the last twenty years has done more harm than good for the environment, and humanity. It's certainly not saving the planet. Not only has it cost trillions of dollars (which could have been spent on reducing pollution and many other worthwhile causes) and failed to reduce CO_2 emissions, but in many cases it has actually increased pollution and reduced people's standard of living and quality of life in various parts of the world.

As for its effect on humanity, convincing a generation of children that they have no future – that the world as we know it is coming to an end within the next couple of decades because of the impending 'climate crisis' – can't be good for their mental health. Child suicide rates, depression, anxiety, and even schoolyard mass shootings by nihilistic children who hate their fellow humans may all be driven to some extent by this sort of doomsday rhetoric.

There are many examples of recent attempts to reduce CO_2 emissions which have done more harm than good.

Promoting the use of biofuels (such as ethanol produced from sugarcane or corn) as an environmentally sustainable alternative to oil is one example. It forced up essential food prices in many countries and used up huge amounts of water, arable land, and chemical fertilizers. Whether or not it actually reduces CO_2 emissions is debatable. And many scientists suspect that runoff from sugarcane farms on Australia's north coast is probably *the* major threat to the Great Barrier Reef anyway, not climate change.

There are numerous scientific studies claiming that forests in the northern hemisphere are considerably more plentiful than a hundred years ago. Part of the reason is that humans started burning fossil fuels instead of trees for cooking and heating. For the past thirty years the 'sustainability' industry has been promoting a reversal of this practice in the belief that because trees absorb CO_2 while growing, they are (supposedly) 'carbon neutral'. But like everything to do with the world's climate and energy and carbon emissions, it's just not that simple. Burning wood (now referred to as biomass) does actually emit CO_2.

It's the type of sleight of hand often used by the climate change movement. Reality is vastly different:

"Reducing the rotation rate of trees to boost forest output is also generally a bad idea for the climate and can lead to the production of biomass that's higher carbon than coal. Several other DECC scenarios also show that burning wood in UK power plants can be worse than coal. This includes harvesting wood from naturally regenerating forest, with emissions of up to 5,174 kilograms per megawatt hour – a staggering five times that of coal." [www.carbonbrief.org]

The obsession with CO_2 emissions also allowed car makers and government agencies to promote the idea that they were helping to save the planet by switching to diesel-powered vehicles because diesel has lower per km CO_2 emissions than petrol. Everyone ignored the obvious – diesel has far higher actual pollution emissions (you can even see and smell it) than petrol, which resulted in cities like London and Oxford ending up with such poor air quality that diesel emissions are causing an estimated 40,000 deaths per year in the UK.

It seems technically possible, and likely, that there'll be a mass conversion to electric vehicles over the next twenty years. Which makes sense because with fewer moving parts and no direct emissions, they are capable of being simpler, cleaner, quieter, faster, and easier to maintain, plus considerably more energy efficient than petrol or diesel-powered vehicles. This would be of enormous benefit in improving air quality in cities around the world.

But the electricity consumption of each household and business is going to increase dramatically, and the effect of an electrified transport fleet on CO_2 emissions depends very much on how that electricity is generated.

An average all-electric grid-connected house in Australia uses at least 20kWh per day. A modern electric car can require as much as 100kWh for a full overnight recharge. This means that many multi-car households will possibly quadruple their electricity consumption. Which is a massive increase.

Power grids, already struggling due to a relatively small input from intermittent renewables, will probably have to at least triple their output, especially at night. This simply can't happen if we don't have an abundant reliable supply of affordable grid electricity, which, as we've discovered, can't be provided by wind turbines and solar panels.

In other words, cities that attempt to power their grid with renewables will make it virtually impossible to convert their transport fleet to electric vehicles.

Then of course there's the constant demand from environmentalists that we shouldn't fly. Judging by the steadily increasing number of airline passengers every year, very few people actually take that to heart (especially all the IPCC delegates constantly flying around the world to conferences), so all it seems to have achieved is to make people feel guilty.

However, if their campaign was actually successful, and passenger numbers dropped by say 20%, that would decimate the airline industry. And the first thing the airlines would do is cancel orders for those new fuel-efficient planes they like to order every few years. We'd be flying around in older, less safe, more polluting planes for the foreseeable future.

And I haven't even mentioned the enormous amounts of steel, aluminium, concrete, and other materials (and energy) required to make all the solar panels, wind turbines, and batteries fuelling this multi-trillion-dollar industry. Let alone the toxic waste disposal required when they're all obsolete in about fifteen years.

These examples indicate a constant theme of climate change activists – an obvious unwillingness or inability to properly investigate what they're doing and think things through. Which once again makes me suspect that what they claim to be hoping to achieve is not their real aim at all.

And I'm beginning to understand why so called 'deniers' often sound so frustrated. It's infuriating to see such poorly thought-out solutions to a problem that increasingly looks as though it doesn't exist – which only make the non-existent problem and the environment even worse – being promoted by people who appear to have very little knowledge of what they're talking about, yet are so certain they are not only right, but also morally superior.

Perhaps if we stopped wasting trillions of dollars every year trying to solve the non-problem of CO_2 emissions, we could focus on the things that really matter. Lifting people out of poverty, improving housing and infrastructure, and cleaning up or preventing further pollution can be expensive endeavours. It's no coincidence that poorer nations are usually the most polluted, polluting, and diseased. And it's worth noting that, in general, democratic countries with an abundant supply of cheap electricity tend to eventually end up with the highest living standards, and the healthiest environments.

If you were genuinely interested in reducing CO_2 emissions and caring for the environment, surely you'd be very interested in measuring or analysing the results of the last twenty years of climate change action. And surely if they weren't improving things, you wouldn't just keep doing more of the same in the forlorn hope that you'll get a different result.

Unless, of course, your real goal is something else entirely.

8. BELIEVE WHAT YOU WILL

*"The mark of your ignorance is the depth of
your belief in injustice and tragedy.
What the caterpillar calls the end of the world,
the Master calls the butterfly."*
Richard Bach

So yes, the climate is changing. And yes, the earth is most probably getting warmer. But only dramatically or catastrophically if we choose to believe the alarmist's computer models. However, as the older and perhaps wiser scientists are saying, computer modelling is not science. And the climate is very complex, we don't fully understand all the variables, and we are kidding ourselves if we think we are able to control everything.

But this is where the climate change movement perfectly fits the narrative of a socialist/atheist belief system – that modern scientists (no matter how badly corrupted or ideologically driven) can understand and ultimately control everything, and if something doesn't suit us, the government should do something about it; an unwavering belief that scientists and governments can fix everything. Despite overwhelming evidence to the contrary.

What if, through no fault of our own, we have almost completely lost (or were never allowed to develop) our ability to think rationally and critically, but instead, are easily led by our feelings and an overwhelming desire to be – or at least appear to be – virtuous?

According to many modern philosophers and psychologists, this is exactly what has occurred through our modern education system, which has been dominated by a postmodernist theology since the nineteen fifties.

Maybe we need to acknowledge that the old religions we've recently discarded – which promoted faith, service, and gratitude as the core values of a meaningful life – led to a happier, more honest, better functioning society than the apparently dysfunctional society we seem to be heading towards. Where instead of gratitude for what we have and what we've achieved, we've been taught to look for what's wrong and demand that the government do something about it. A society of easily-led 'social justice warriors' full of self-declared virtue and indignation fuelling our need for meaning; with a lifetime of mental health issues lurking in our fragile future.

Harvard psychologist Steven Pinker points out in *Enlightenment Now; The Case for Reason, Science, Humanism and Progress*, that almost every aspect of human life has improved markedly (except perhaps mental health) in the past couple of hundred years – including life expectancy, violence, war, poverty, inequality, human rights, *and* the environment. Hard to believe if you follow modern mainstream media, but statistical evidence appears to support Pinker's case.

There's always been an apocalyptical fear in the human psyche, and yet somehow we keep muddling our way through and making undeniable progress. But doomsday scenarios and sensationalist news always get attention. A narrative that; *Things aren't perfect, but we're working on it, and everything will probably be okay*, isn't likely to get much attention, but it's generally closer to the truth.

The last great environmental scare was perpetrated by Paul Ehrlich's book *The Population Bomb*, published in 1968, which confidently declared:

"The battle to feed all of humanity is over. In the 1970s hundreds of millions of people will starve to death in spite of any crash programs embarked upon now… conscious regulation of human numbers must be achieved."

A suggested solution to this impending crisis included mass sterilization programs. Of course it was a movement that never really caught on, for many reasons. Including perhaps that there was no money to be made from it, and it was difficult to portray yourself as virtuous by promoting such an obviously anti-humanist ideology. And it turned out to be spectacularly wrong.

Although the world's population has continued to increase, the latest research, as pointed out in Oxford Professor Danny Dorling's (2013) book *Population 10 Billion,* indicates that the rate of increase is rapidly declining and the world's population will most likely level off at ten billion before slowly declining. And perhaps the most surprising thing is that we need do nothing to make this happen apart from what we humans tend to do quite naturally; constantly striving to raise our standard of living throughout the world.

Recent history demonstrates quite clearly that as each country develops economically, living conditions rise, poverty declines, education levels increase, women's participation in the workforce increases, and birth rates drop spectacularly to less than replacement level within one generation. But with the current doomsday predictions of human-induced climate change, we've now got an anti-humanist

148

movement disguised as an environmental movement that not only can people make a lot of money from and build a career out of, but they can also appear virtuous at the same time, being able to claim they're 'saving the planet'.

However, when the basis of a belief turns out to be completely false (the '97% of scientists agree' mantra), you really have to question not only the validity of the belief, but also the intentions and integrity of those advocating it. This perhaps is borne out by the attitude of the IPCC scientists in the leaked emails.

While those who think that the end justifies the means – perhaps a little exaggeration here and there to frighten ignorant, self-centred, short-sighted people into action – history has shown us that in most cases the means *are* the end. The means, or method, are a sure indicator of people's true beliefs and virtue.

And when you question a hard-core environmentalist, you generally uncover an anti-humanist attitude, often driven by nothing more than envy of those who are more accomplished or wealthier. Which is understandable. If you've convinced yourself that CO_2 is going to destroy the planet, how can you possibly not believe that seven billion increasingly wealthy, constantly-consuming, CO_2-emitting humans are undoubtedly its main threat. It's a dangerously nihilistic belief.

Climate change alarmism is more of an atheist post-modernist political movement than a scientifically-based environmental movement. The Green party, which masquerades as the obvious political choice of any serious environmentalist, makes their real intentions clear in their 1989 manifesto:

"Survival is highly motivating and may help us in building a mass movement that will lead to large scale political and societal change in a very short period of time."

They make no secret of their real goal, social justice, which they define as: *"...achieving equality of distribution of goods and wealth amongst all groups in society."* Which is all very well if that's what you believe will lead to a better world. But there are going to be a lot of decent, well-intentioned people who will be disillusioned and angry when they discover that they have been deceived by a political movement led by people far more concerned about *'societal change'* than they are about the environment. A movement led by people so certain they are right that they believe the end justifies the means and will do anything to achieve their aim; deliberately deceiving, bullying, and vilifying others in order to achieve their utopian goals. Because what they claim to be doing – helping the environment/saving the planet – has nothing to do with the true goals of a totalitarian socialist political system, which taken to the extreme, historically, has resulted in the dystopian nightmares of Hitler's Germany, Stalin's Russia, and Mao's China; regimes that also believed the end justified the means.

While a compliant media deceptively portrays this as a battle between 'science believers' and 'science deniers' (which is a classic case of psychological projection), it's really a battle between fashionable pseudoscience and real science, between utopian socialism and pragmatic capitalism, and between those who believe that man is the master of the universe and those who accept that perhaps the *universe* itself (or God, if you're religious) is the master of the universe.

It's also a battle between those who believe that a better world will be created by focusing on redistribution of wealth, and others who believe it's better to focus on creating wealth. But as Dr Richard Lindzen points out in one of his lectures, the inevitable result of climate change action is generally a transfer of wealth from the poor of wealthy countries, to the wealthy of poor countries.

History has clearly demonstrated that democratic countries which have focused on the creation of wealth have always ended up with a greater proportion of its citizens living a higher standard of living than countries focused on wealth distribution, which nearly always fail.

It's dangerous to mislead an innocent public because it destroys trust and tears apart the fabric of society (will we ever trust government institutions, mainstream media, and scientific organisations again?). So it would be good if mainstream media had the integrity and self-awareness to start reporting the truth instead of broadcasting their politically-inspired opinions and ideologies. And perhaps political movements could at least be a little more open and honest about their philosophies and intentions, so if enough people agreed with their aims and values they may even get legitimately voted into power. That's the way it usually works in a democracy.

After what I've discovered about climate science in the past few months, it's been a humbling experience to challenge my beliefs and opinions and eventually change my mind. Actually, it's embarrassing to admit how badly informed I was about something so important and relevant to my own career. Obviously it's not in my own interest to come to the conclusion I've settled on. It certainly doesn't fit my self-image as a well-informed, compassionate environmentalist; because as we've been told, repeatedly, for many years; anyone who doesn't believe in climate change is an ignorant, uneducated, uncaring, selfish bigot. Aren't they?

So now for the hard part. If you've read this far at least you're curious and open minded. But my words are easily dismissed as just another opinion on a complex and confusing issue, and you probably won't watch any of the YouTube presentations of the scientists I've listed. Because you really don't *want* to have your mind changed. After all, who wants to be a 'denier'? What would your friends think?

This is the most insidious thing about the climate change movement. Somehow they've managed – with plenty of help from mainstream media and virtue-signalling celebrities and politicians – to frame the situation so that if you don't believe that the majority of your fellow human beings are greedy, self-centred creatures who are destroying the planet by carelessly burning fossil fuels, then you must be an ignorant, immoral fool who shouldn't even be allowed to express an opinion, no matter how qualified or well informed you may be. And this is where many modern environmentalists need to be called out for what they are.

Living a life where virtually every single aspect of our existence is dependent on and provided by a fossil-fuel-powered capitalist industrial complex, while complaining about and denigrating the intelligence and morality of the people who are burdened with the task of making it all work, is surely the height of ignorance and hypocrisy. Inner city Greens protest and mock the 'deplorables' who build the houses, smartphones, coffee machines, and bicycles they can't live without; while living a life of relative luxury. And yes, we *are* living a life of luxury if we don't have to labour all day on a hot building site, or in a factory, or down a mine shaft, or out in the wind and rain in the middle of the night working to keep the power on.

Only someone who knows absolutely nothing about how things are made, and how the world works, would believe they could live without using anything that wasn't made, transported, processed, grown, or packaged without some sort of fossil fuel. Steel, for instance, can't be made without coal – and without steel there would be no buildings, no cars, no planes, no trains, and no coffee machines?

The truth is, that if the science of climate change was a thousand-page book, the theory that it is caused predominantly by burning fossil fuels would be one page of that book. Tearing that page out and waving it around claiming that everyone who doesn't agree is a science-denier is not helpful, or honest.

Prince Harry and his new bride recently visited Australia, and of course entrenched his popularity with the public and media by listing recent droughts, coral bleaching, and severe storms as evidence for the need of immediate climate action.

I doubt that the prince has much knowledge of climate science, but even if he did, he's not stupid. Imagine the outrage and ridicule he would have faced had he said something like; *Well, actually, I've spent months looking carefully into the climate change debate and have discovered that most of the more experienced scientists whose careers, income, and reputation aren't dependent on perpetuating the belief in climate change alarmism, actually don't think there's any evidence to support the need for a drastic reduction in the use of fossil fuels.* Harry would have been hounded by the media and ridiculed for the rest of his life.

What a ridiculous situation we've found ourselves in. It's going to take a lot of people with a lot of courage to get the truth out. And it's going to be embarrassingly difficult for some scientists and many politicians and public intellectuals to back away from this with some dignity. They have staked their whole reputation and career on a simplistic hypothesis that was widely accepted and adopted before being adequately tested, and too many people desperately wanted to be true for reasons that have nothing to do with the environment.

It's hard not to be disillusioned and angry when you first realise that you've been badly deceived, and have believed in a lie for many years. But after thinking it through, if you're not absolutely thrilled to learn that we're not destroying the planet with CO_2 after all, you probably need to question your own desires and motives. They can't be good.

Think about it. For many of us, most of our lives we've believed that just by living a normal life – consuming, driving, watching TV, flying, eating, defecating, even breathing – we're helping to destroy the planet by emitting carbon dioxide. That's a hell of a burden. And it's not true.

It's naïve to not at least consider that those trying to make us feel scared, guilty, and helpless may actually be trying to manipulate us. And after what I've learnt in the last six months, it seems obvious to me that there's a political/ideological movement that is deliberately using misinformation to keep people in a constant state of confusion, guilt, fear, outrage, and helplessness.

It's a movement led by those who believe that humans need to be manipulated, socially engineered, frightened, or shamed (or even deceived if necessary) into behaving decently – with a never-ending cascade of laws controlling their every word and action. People who are either misinformed or uninformed – but are fundamentally decent and compassionate, with a strong desire to be virtuous – are easily led by the deceptions of this movement.

But perhaps the reality is that most people are fundamentally good, and given the opportunity and freedom, and told the truth, will generally do the right thing; not only for themselves and those close to them, but also for society at large, without the need of a myriad of draconian laws to keep them under control. There's plenty of evidence to suggest that there's an inherent decency in most, if not all, people, and the more advanced and civilized a society is, the fewer laws it tends to need.

As Steven Pinker declares realistically in his book *Enlightenment Now*: *"We will never have a perfect world, but – defying the chorus of fatalism and reaction – we can continue to make it a better one."*

That's what I believe.

9. CONCLUSION

It's quite bizarre that a quasi-religious political movement has managed to demonize carbon dioxide to the extent that a large proportion of the population actually believe that an invisible, odourless, natural gas that makes up a tiny percentage of the earth's atmosphere is catastrophically changing the world's climate. Especially when – if anything – the increase in atmospheric CO_2 over the past century has helped revegetate the earth. Which may well be nature's way of accommodating an increasing human population. If you have faith in a benevolent creator, then that's what you're inclined to believe.

Perhaps it's even more bizarre that so many people with virtually no knowledge of science, history, or the way things are made and work, are so convinced that anyone who disagrees with their TV-induced beliefs are not only denying science, but also denying reality. Surely that's psychological projection at its very best.

Evidence that the earth is warming is not evidence that the warming is caused by burning fossil fuels. Thirty years ago the CO_2-driven global warming theory was a theory worth investigating, but it turned out to be a massive oversimplification of a very complex system we barely understand, let alone control. In fact it's quite clear that burning fossil fuels in the past hundred years has had no significant influence on the global climate or sea levels.

And the climate change movement is a political movement supported by a predominantly well-intentioned but naive and scientifically ignorant portion of the world's population. It is not an environmental movement. Its most avid proponents know nothing about science and care little about the environment; even less about people.

So the good news is that the planet will be just fine; especially if we acknowledge the truth and ignore the distractions of the climate change movement and their obsession with trying to reduce emissions of a harmless gas. In which case we can return to the sort of real environmentalism that existed well before the advent of climate alarmism; focusing instead on eliminating dangerous pollutants, effective waste disposal, protecting natural ecosystems, and revegetating as much of the planet as possible.

But we need to beware.

It may seem, at first, a bit over-the-top to describe the climate change movement as a dangerous belief system that could easily morph into a truly evil totalitarian political movement. But I really do think it could be that serious. A core belief of many people at the moment is that there are simply too many of us, and we are destroying the planet.

It usually goes unsaid that – perversely – they believe humanity can only be saved by a massive reduction in population and our immoral consumptive way of living. In other words, we have to wipe out a lot of humanity and drastically lower our standard of living in order to save humanity, or more importantly, the planet.

What they really believe, apparently, is that humanity is a cancer, or plague, on the Earth, which would be a lot better off without us – a view openly expressed by respected celebrity environmentalists such as David Attenborough and parroted by clueless kids at Extinction Rebellion rallies. According to Canadian psychologist Jordan Peterson; *"You can't utter a more genocidal phrase than that ... and of course you do it in a display of your care for the world. It's really sickening, it's appalling. There's a hatred for humanity at the bottom of it."*

One young protester I talked to recently assured me with an eerie mixture of anger, glee, and absolute certainty that the Earth would soon *"rid itself of humans like a dog shaking off its fleas"*.

That sent a shiver down my spine.

To extend an earlier analogy, there's an elephant in your backyard. Yes, I know, all your friends and trusted authorities tell you otherwise, and I'm just an unknown layman. So rant and rave if you like, do your best to discredit me, my research methods, my intelligence, my motives. Find some little fact or reference I've got wrong in order to discredit my whole book. I don't care. But at least poke your head out the back door and have a look for yourself.

You might be surprised by what you find.

I think it's worth concluding this book with the words and wisdom of someone far more eloquent, revered, and worldly than myself. Celebrated Australian journalist Clive James – apparently freed by his impending death from the fear of upsetting his politically-correct literary fans and colleagues – penned the following essay in 2019:

"When you tell people once too often that the missing extra heat is hiding in the ocean, they will switch over to watch Game of Thrones, where the dialogue is less ridiculous and all the threats come true. The proponents of man-made climate catastrophe asked us for so many leaps of faith that they were bound to run out of credibility in the end.

Now that they finally seem to be doing so, it could be a good time for those of us who have never been convinced by all those urgent warnings to start warning each other that we might be making a comparably senseless tactical error if we expect the elastic cause of the catastrophists, and all of its exponents, to go away in a hurry.

I speak as one who knows nothing about the mathematics involved in modelling non-linear systems. But I do know quite a lot about the mass media, and far too much about the abuse of language.

So I feel qualified to advise against any triumphalist urge to compare the apparently imminent disintegration of the alarmist cause to the collapse of a house of cards. Devotees of that fond idea haven't thought hard enough about their metaphor. A house of cards collapses only with a sigh, and when it has finished collapsing all the cards are still there.

Although the alarmists might finally have to face that they will not get much more of what they want on a policy level, they will surely, on the level of their own employment, go on wanting their salaries and prestige. To take a conspicuous if ludicrous case, the Australian climate star Tim Flannery will probably not, of his own free will, shrink back to the position conferred by his original metier, as an expert on the extinction of the giant wombat. He is far more likely to go on being, and wishing to be, one of the mass media's mobile oracles about climate. While that possibility continues, it will go on being dangerous to stand between him and a TV camera. If the giant wombat could have moved at that speed, it would still be with us.

The mere fact that few of Flannery's predictions have ever come even remotely true need not be enough to discredit him. The same fact, in the case of America's Professor Ehrlich, has left him untouched ever since he predicted that the world would soon run out of copper. In those days, when our current phase of the long discussion about man's attack on nature was just beginning, he predicted mass death by extreme cold. Lately he predicts mass death by extreme heat. But he has always predicted mass death by extreme something, and he is always Professor Ehrlich.

Actually, a more illustrative starting point for the theme of the permanently imminent climatic apocalypse might be taken as 3 August 1971, when the Sydney Morning Herald announced that the Great Barrier Reef would be dead in six months. After six months the reef had not died, but it has been going to die almost as soon as that ever since; making it a strangely durable emblem for all those who have wedded themselves to the notion of climate catastrophe.

The most exalted of all the world's predictors of reef death, President Obama, has still not seen the reef even now but he promises to go there one day when it is well again. Assurances that it has never really been sick won't be coming from his senior science adviser John Holdren. In the middle of 2016 some of the long-term experts on reef death began admitting that they had all been overdoing the propaganda. After almost half a century of reef death prediction, this was the first instance of one group of reef death predictors telling another group to dial down the alarmism, or they would queer the pitch for everybody.

But an old hand like Holdren knows better than to listen to sudden outbursts of moderation. Back in the day, when extreme cooling was the fashion, he was an extreme coolist. Lately he is an extreme warmist. He will surely continue to be an extremist of some kind, even if he has to be an extreme moderate. And after all, his boss was right about the ocean. In his acceptance speech at the 2008 Democratic convention, Obama said – and I truly wish that this were an inaccurate paraphrase – that people should vote for him if they wanted to stop the ocean rising. He got elected, and it didn't rise.

The notion of a count-down or a tipping point is very dear to both wings of this deaf shouting match, and really is of small use to either. On the catastrophist wing, whose 'narrative', as they might put it, would so often seem to be a synthesised film script left over from the era of surround-sound disaster movies, there is always a count-down to the tipping point. When the scientists are the main contributors to the script, the tipping point will be something like the forever forthcoming moment when the Gulf Stream turns upside down or the Antarctic ice sheet comes off its hinges, or any other extreme event which, although it persists in not happening, could happen sooner than we think (science correspondents who can write a phrase like 'sooner than we think' seldom realise that they might have already lost you with the word 'could').

When the politicians join in the writing, the dramatic language declines to the infantile. There are only 50 days (Gordon Brown) or 100 months (Prince Charles wearing his political hat) left for mankind to 'do something' about 'the greatest moral challenge … of our generation'. (Kevin Rudd, before he arrived at the Copenhagen climate shindig in 2009.)

When he left Copenhagen, Rudd scarcely mentioned the greatest moral challenge again. Perhaps he had deduced, from the confusion prevailing throughout the conference, that the chances of the world ever uniting its efforts to 'do something' were very small. Whatever his motives for backing out of the climate chorus, his subsequent career was an early demonstration that to cease being a chorister would be no easy retreat, because it would be a clear indication that everything you had said on the subject up to then had been said in either bad faith or ignorance. It would not be enough merely to fall silent. You would have to travel back in time, run for office in the Czech Republic instead of Australia, and call yourself Vaclav Klaus.

Australia, unlike Kevin Rudd, has a globally popular role in the climate movie because it looks the part. Common reason might tell you that a country whose contribution to the world's emissions is only 1.4% can do very little about the biggest moral challenge even if it manages to reduce that contribution to zero; but your eyes tell you that Australia is burning up. On the classic alarmist principle of 'just stick your head out of the window and look around you', Australia always looks like Overwhelming Evidence that the alarmists must be right.

Even now that the global warming scare has completed its transformation into the climate change scare so that any kind of event at either end of the scale of temperature can qualify as a crisis, Australia remains the top area of interest, still up there ahead of even the melting North Pole, despite the Arctic's miraculous capacity to go on producing ice in defiance of all instructions from Al Gore. A 'C' student to his marrow, and thus never quick to pick up any reading matter at all, Gore has evidently never seen the Life magazine photographs of America's nuclear submarine Skate surfacing through the North Pole in 1959. The ice up there is often thin, and sometimes vanishes. But it comes back, especially when someone sufficiently illustrious confidently predicts that it will go away for good.

After 4.5 billion years of changing, the climate that made outback Australia ready for Baz Luhrmann's view-finder looked all set to end the world tomorrow. History has already forgotten that the schedule for one of the big drought sequences in his movie Australia was wrecked by rain, and certainly history will never be reminded by the mass media, which loves a picture that fits the story.

In this way, the polar bear balancing on the photo-shopped shrinking ice-floe will always have a future in show business, and the cooling towers spilling steam will always be up there in the background of the TV picture while the panel of experts discuss what Julia Gillard still calls 'carbon', her word for carbon dioxide. Pictures of her house near the beach in Adelaide, on the other hand, will never be used to illustrate satirical articles about a retired prophet of the rising ocean who buys a house near the beach, because there won't be any such articles.

The full 97% of all satirists who dealt themselves out of the climate subject back at the start look like staying out of it until the end, even if they get satirised in their turn. One could blame them for their pusillanimity, but it would be useless, and perhaps unfair. Nobody will be able plausibly to call Emma Thompson dumb for spreading gloom and doom about the climate: she's too clever and too creative. And anyway, she might be right. Cases like Leonardo di Caprio and Cate Blanchett are rare enough to be called brave.

Otherwise, the consensus of silence from the wits and thespians continues to be impressive. If they did wish to speak up for scepticism, however, they wouldn't find it easy when the people who run the big TV outlets forbid the wrong kind of humour. On Saturday Night Live back there in 2007, Will Ferrell, brilliantly pretending to be George W. Bush, was allowed to get every word of the global warming message wrong, but he wasn't allowed to disbelieve it.

Just as all branches of the modern media love a picture of something that might be part of the Overwhelming Evidence for climate change even if it is really a picture of something else, they all love a clock ticking down to zero, and if the clock never quite gets there then the motif can be exploited forever.

But the editors and producers must face the drawback of such perpetual excitement: it gets perpetually less exciting. Numbness sets in, and there is time to think after all. Some of the customers might even start asking where this language of rubber numbers has been heard before.

It was heard from Swift. In Gulliver's Travels he populated his flying island of Laputa with scientists busily using rubber numbers to predict dire events. He called these scientists 'projectors'. At the basis of all the predictions of the projectors was the prediction that the Earth was in danger from a Great Comet whose tail was 'ten hundred thousand and fourteen' miles long.

I should concede at this point that a sardonic parody is not necessarily pertinent just because it is funny; and that although it might be unlikely that the Earth will soon be threatened by man-made climate change, it might be less unlikely that the Earth will be threatened eventually by an asteroid, or let it be a Great Comet; after all, the Earth has been hit before. That being said, however, we can note that Swift has got the language of artificial crisis exactly right, to the point that we might have trouble deciding whether he invented it, or merely copied it from scientific voices surrounding him in his day. James

Hansen is a Swiftian figure. Blithely equating trains full of coal to trains full of people on their way to Auschwitz, Hansen is utterly unaware that he has not only turned the stomachs of the informed audience he was out to impress, he has lost their attention.

Professor of Earth Sciences Chris Turney, who led a ship full of climate change enthusiasts into the Antarctic ice to see how the ice was doing under the influence of climate change and found it was doing well enough to trap the ship, could have been invented by Swift. (Turney's subsequent Guardian article, in which he explained how this embarrassment was due only to a quirk of the weather, and had nothing to do with a possible mistake about the climate, was a Swiftian lampoon in all respects.)

Compulsorily retired now from the climate scene, Dr Rajendra Pachauri was a zany straight from Swift, by way of a Bollywood remake of The Party starring the local imitator of Peter Sellers; if Dr Johnson could have thought of Pachauri, Rasselas would be much more entertaining than it is. Finally, and supremely, Tim Flannery could have been invented by Swift after ten cups of coffee too many with Stella. He wanted to keep her laughing. Swift projected the projectors who now surround us.

They came out of the grant-hungry fringe of semi-science to infect the heart of the mass media, where a whole generation of commentators taught each other to speak and write a hyperbolic doom-language ('unprecedented', 'irreversible', etcetera), which you might have thought was sure to doom them in their turn. After all, nobody with an intact pair of ears really listens for long to anyone who talks about 'the planet' or 'carbon' or 'climate denial' or 'the science'.

But for now – and it could be a long now – the advocates of drastic action are still armed with a theory that no fact doesn't fit. The theory has always been manifestly unfalsifiable, but there are few science pundits in the mass media who could tell Karl Popper from Mary Poppins.

More startling than their ignorance, however, is their defiance of logic. You can just about see how a bunch of grant-dependent climate scientists might go on saying that there was never a Medieval Warm Period even after it has been pointed out to them that any old corpse dug up from the permafrost could never have been buried in it. But how can a bunch of supposedly enlightened writers go on saying that? Their answer, if pressed, is usually to say that the question is too elementary to be considered.

Alarmists have always profited from their insistence that climate change is such a complex issue that no 'science denier' can have an opinion about it worth hearing. For most areas of science such an insistence would be true. But this particular area has a knack of raising questions that get more and more complicated in the absence of an answer to the elementary ones.

One of those elementary questions is about how man-made carbon dioxide can be a driver of climate change if the global temperature has not gone up by much over the last twenty years but the amount of man-made carbon dioxide has. If we go on to ask a supplementary question – say, how could carbon dioxide raise temperature when the evidence of the ice cores indicates that temperature has always raised carbon dioxide – we will be given complicated answers, but we still haven't had an answer to the first question, except for the suggestion that the temperature, despite the observations, really has gone up, but that the extra heat is hiding in the ocean.

It is not necessarily science denial to propose that this long professional habit of postponing an answer to the first and most elementary question is bizarre. Richard Feynman said that if a fact doesn't fit the theory, the theory has to go. Feynman was a scientist. Einstein realised that the Michelson-Morley experiments hinted at a possible fact that might not fit Newton's theory of celestial mechanics. Einstein was a scientist too.

Those of us who are not scientists, but who are sceptical about the validity of this whole issue – who suspect that the alleged problem might be less of a problem than is made out – have plenty of great scientific names to point to for exemplars, and it could even be said that we could point to the whole of science itself. Being resistant to the force of its own inertia is one of the things that science does.

When the climatologists upgraded their frame of certainty from global warming to climate change, the bet-hedging manoeuvre was so blatant that some of the sceptics started predicting in their turn; the alarmist cause must surely now collapse, like a house of cards. A tipping point had been reached.

Unfortunately for the cause of rational critical enquiry, the campaign for immediate action against climate doom reaches a tipping point every few minutes, because the observations, if not the calculations, never cease exposing it as a fantasy.

I myself, after I observed Andrew Neil on BBC TV wiping the floor with the then Secretary for Energy and Climate Change Ed Davey, thought that the British government's energy policy could not survive, and that the mad work which had begun with Ed Miliband's Climate Act of 2008 must now surely begin to come undone. Neil's well-informed list of questions had been a tipping point. But it changed nothing in the short term. It didn't even change the BBC, which continued uninterrupted with its determination that the alarmist view should not be questioned.

How did the upmarket mass media get themselves into such a condition of servility? One is reminded of that fine old historian George Grote, when he said that he had taken his A History of Greece only to the point where the Greeks themselves failed to realise they were slaves. The BBC's monotonous plugging of the climate theme in its science documentaries is too obvious to need remarking, but it's what the science programmes never say that really does the damage.

Even the news programmes get 'smoothed' to ensure that nothing interferes with the constant business of protecting the climate change theme's dogmatic status. To take a simple but telling example: when Sigmar Gabriel, Germany's Vice Chancellor and man in charge of the Energiewende, talked rings around Greenpeace hecklers with nothing on their minds but renouncing coal, or told executives of the renewable energy companies that they could no longer take unlimited subsides for granted, these instructive moments could be seen on German television but were not excerpted and subtitled for British television even briefly, despite Gabriel's accomplishments as a natural TV star, and despite the fact that he himself was no sceptic.

Wrong message: easier to leave him out. And if the climate scientist Judith Curry appears before a US Senate committee and manages to defend her anti-alarmist position against concentrated harassment from a senator whose only qualification for the discussion is that he can impugn her integrity with a rhetorical contempt of which she is too polite to be capable? Leave it to YouTube. In this way, the BBC has spent ten years unplugged from a vital part of the global intellectual discussion, with an increasing air of provincialism as the inevitable result.

As the UK now begins the long process of exiting the European Union, we can reflect that the departing nation's most important broadcasting institution has been behaving, for several years, as if its true aim were to reproduce the thought control that prevailed in the Soviet Union. As for the print media, it's no mystery why the upmarket newspapers do an even more thorough job than the downmarket newspapers of suppressing any dissenting opinion on the climate.

In Britain, the Telegraph sensibly gives a column to the diligently sceptical Christopher Booker, and Matt Ridley has recently been able to get a few rational articles into the Times, but a more usual arrangement is exemplified by my own newspaper, the Guardian, which entrusts all aspects of the subject to George Monbiot, who once informed his green readership that there was only one reason I could presume to disagree with him, and them: I was an old man, soon to be dead, and thus with no concern for the future of 'the planet'. I would have damned his impertinence, but it would have been like getting annoyed with a wheelbarrow full of freshly cut grass.

These byline names are stars committed to their opinion, but what's missing from the posh press is the non-star name committed to the job of building a fact-file and extracting a reasoned article from it. Further down the market, when the Daily Mail put its no-frills news-hound David Rose on the case after Climategate, his admirable competence immediately got him labelled as a 'climate change denier': one of the first people to be awarded that badge of honour.

The other tactic used to discredit him was the standard one of calling his paper a disreputable publication. It might be – having been a victim of its prurience myself, I have no inclination to revere it – but it hasn't forgotten what objective reporting is supposed to be. Most of the British papers have, and the reason is no mystery.

They can't afford to remember. The print media are on their way down the drain. With almost no personnel left to do the writing, the urge at editorial level is to give all the science stuff to one bloke. The print edition of The Independent bored its way out of business when their resident climate nag was allowed to write half the paper.

In its last year, when the doomwatch journalists were threatened by the climate industry with a newly revised consensus opinion that a mere two-degree increase in world temperature might be not only acceptable but likely, the Independent's chap retaliated by writing stories about how the real likelihood was an increase of five degrees, and in a kind of frenzied crescendo he wrote a whole front page saying that the global temperature was 'on track' for an increase of six degrees. Not long after, the Indy's print edition closed down.

At the New York Times, Andrew Revkin, star colour-piece writer on the climate beat, makes the whole subject no less predictable than his prose style: a cruel restriction. In Australia, the Fairfax papers, which by now have almost as few writers as readers, reprint Revkin's summaries as if they were the voice of authority, and will probably go on doing so until the waters close overhead.

On the ABC, the house science pundit Robyn Williams famously predicted that the rising of the waters 'could' amount to 100 metres in the next century. But not even he predicted that it could happen next week. At the Sydney Morning Herald, it could happen next week. The only remaining journalists could look out of the window, and see fish.

Bending their efforts to sensationalise the news on a scale previously unknown even in their scrappy history, the mass media have helped to consolidate a pernicious myth. But they could not have done this so thoroughly without the accident that they are the main source of information and opinion for people in the academic world and in the scientific institutions. Few of those people have been reading the sceptical blogs: they have no time.

If I myself had not been so ill during the relevant time-span, I might not have been reading them either, and might have remained confined within the misinformation system where any assertion of forthcoming disaster counts as evidence. The effect of this mountainous accumulation of sanctified alarmism on the academic world is another subject.

Some of the universities deserve to be closed down, but I expect they will muddle through, if only because the liberal spirit, when it regains its strength, is likely to be less vengeful than the dogmatists were when they ruled. Finding that the power of inertia blesses their security as once it blessed their influence, the enthusiasts might have the sense to throttle back on their certitude, huddle under the blanket cover provided by the concept of 'post-normal science', and wait in comfort to be forgotten.

As for the learned societies and professional institutions, it was never a puzzle that so many of them became instruments of obfuscation instead of enlightenment. Totalitarianism takes over a state at the moment when the ruling party is taken over by its secretariat; the tipping point is when Stalin, with his lists of names, offers to stay late after the meeting and take care of business. The same vulnerability applies to any learned institution. Rule by bureaucracy favours mediocrity, and in no time at all you are in a world where Julia Slingo is a figure of authority, and Judith Curry is fighting to breathe. Under Stalin, Trofim Lysenko became more indispensable the more he reduced all the other biologists to the same condition as Soviet agriculture, and even after Stalin was dead, it took Andrei Sakharov to persuade Khrushchev not to bring Lysenko back to office. Khrushchev was well aware that Lysenko was a charlatan, but he looked like an historic force; and who argues with one of those?

On a smaller scale of influential prestige, Lord Stern lends the Royal Society the honour of his presence. For those of us who regard him as a vocalised stuffed shirt, it is no use saying that his confident pronouncements about the future are only those of an economist.

Vaclav Klaus was only an economist when he tried to remind us that Malthusian clairvoyance is invariably a harbinger of totalitarianism. But Klaus was a true figure of authority. Alas, true figures of authority are in short supply, and tend not to have much influence when they get to speak. All too often, this is because they care more about science than about the media.

As recently as 2015, after a full ten years of nightly proof that this particular scientific dispute was a media event before it was anything, Freeman Dyson was persuaded to go on television. He was up there just long enough to say that the small proportion of carbon dioxide that was man-made could only add to the world's supply of plant food.

The world's mass media outlets ignored the footage, mainly because they didn't know who he was. I might not have known either if I hadn't spent, in these last few years, enough time in hospitals to have it proved to me on a personal basis that real science is as indispensable for modern medicine as cheap power. Among his many achievements, to none of which he has ever cared about drawing attention, Dyson designed the TRIGA reactor. The TRIGA ensures that the world's hospitals get a reliable supply of isotopes.

Dyson served science. Except for the few hold-outs who go on fighting to defend the objective nature of truth, most of the climate scientists who get famous are serving themselves. There was a time when the journalists could have pointed out the difference, but now they have no idea. Instead, they are so celebrity-conscious that they would supply Tim Flannery with a new clown-suit if he wore out the one he is wearing now. In 2016, he dived on the Great Barrier Reef and reported himself overwhelmed by the evidence that it was on the point of death, a symptomatology which, he said, he had recently learned to recognise by watching his father die. Neither he nor any of his admirers at the Sydney Morning Herald cared to note that it has now been almost 50 years that the reef has been going to die soon. But the moment never came, although it will probably go on being about to happen for the next 50 years as well.

The reef death disaster is like those millions of climate change refugees who were going to flood into the West by 2010. They never arrived. But when the refugees from the war in Syria started to arrive, there was a ready-made media apparatus waiting to declare that they were the missing climate change refugees really, because what else had caused the war but climate change? They were the missing heat that had been hiding in the ocean.

A bad era for science has been a worse one for the mass media, the field in which, despite the usual blunders and misjudgements, I was once proud to earn my living. But I have spent too much time, in these last few years, being ashamed of my profession: hence the note of anger which, I can now see, has crept into this essay even though I was determined to keep it out. As my retirement changed to illness and then to dotage, I would have preferred to sit back and write poems than to be known for taking a position in what is, despite the colossal scale of its foolish waste, a very petty quarrel. But when some of the climate priesthood, and even the Attorney General of the United States, started talking about how dissent might be suppressed with the force of law – well, that was a tipping point. I am a dissenter, and not because I deny science, but because I affirm it.

So it was time to stand up and fight, if only because so many of the advocates, though they must know by now that they are professing a belief they no longer hold, will continue to profess it anyway.

Back in the day, when I was starting off in journalism – on the Sydney Morning Herald, as it happens – the one thing we all learned early from our veteran colleagues was never to improve the truth for the sake of the story. If they caught us doing so, it was the end of the world.

But here we are, and the world hasn't ended after all. Though some governments might not yet have fully returned to the principle of evidence-based policy, most of them have learned to be wary of policy-based evidence. They have learned to spot it coming, not because the real virtues of critical enquiry have been well argued by scientists, but because the false claims of abracadabra have been asserted too often by people who, though they might have started out as scientists of a kind, have found their true purpose in life as ideologists.

Modern history since World War II has shown us that it is unwise to predict what will happen to ideologists after their citadel of power has been brought low. It was feared that the remaining Nazis would fight on, as Werewolves. Actually, only a few days had to pass before there were no Nazis to be found anywhere except in Argentina, boring one another to death at the world's worst dinner parties.

After the collapse of the Soviet Union, on the other hand, when it was thought that no apologists for Marxist collectivism could possibly keep their credibility in the universities of the West, they not only failed to lose heart, they gained strength. Some critics would say that the climate change fad itself is an offshoot of this lingering revolutionary animus against liberal democracy, and that the true purpose of the climatologists is to bring about a world government that will ensure what no less a philanthropist than Robert Mugabe calls 'climate justice', in which capitalism is replaced by something more altruistic.

I myself prefer to blame mankind's inherent capacity for raising opportunism to a principle: the enabling condition for fascism in all its varieties, and often an imperative mind-set among high end frauds. On behalf of the UN, Maurice Strong, the first man to raise big money for climate justice, found slightly under a million dollars of it sticking to his fingers, and hid out in China for the rest of his life – a clear sign of his guilty knowledge that he had pinched it.

Later operators lack even the guilt. They just collect the money, like the Prime Minister of Tuvalu, who has probably guessed by now that the sea isn't going to rise by so much as an inch; but he still wants, for his supposedly threatened atoll, a share of the free cash, and especially because the question has changed. It used to be: how will we cope when the disaster comes? The question now is: how will we cope if it does not?

There is no need to entertain visions of a vast, old-style army of disoccupied experts retreating through the snow, eating first their horses and finally each other. But there could be quite a lot of previously well-subsidised people left standing around while they vaguely wonder why nobody is listening to them anymore.

Way back there in 2011, one of the Climategate scientists, Tommy Wils, with an engagingly honest caution rare among prophets, speculated in an email about what people outside their network might do to them if climate change turned out to be a bunch of natural variations: 'Kill us, probably.' But there has been too much talk of mass death already, and anyway most of the alarmists are the kind of people for whom it is a sufficiently fatal punishment simply to be ignored."

[This is an extract from the essay *Mass Death Dies Hard* by Clive James in *Climate Change: The Facts* 2017, edited by Jennifer Marohasy, published by the Institute of Public Affairs]

ACKNOWLEDGEMENTS

As someone who is self-employed, financially secure, and has no public profile, it's relatively easy for me to make a stance against the socially-acceptable politically-correct position of climate change alarmism. I might be shunned by some people I know and meet who happen to be true 'believers' and are offended by anyone with a different opinion. But that's no great sacrifice for me. However, there are many scientists, writers, and commentators who have made some serious sacrifices – to their reputation, personal lives, and careers – in order to tell the truth. I thank them all for their integrity and courage.

REFERENCES

WEBSITES

www.lavoisier.com.au
www.undeceivingourselves.org/I-ipcc.htm
www.heartland.org
www.wattsupwiththat.com
www.drroyspencer.com
www.defyccc.com/heidelberg-appeal-anniversary
www.euanmearns.com

YOUTUBE LECTURES & INTERVIEWS

Dr Don Easterbrook
Dr Jennifer Marohasy
Dr Bob Carter
Dr John Christy
Dr William Happer
Dr Freeman Dyson
Dr Nir Shaviv
Dr Judith Curry
Dr Timothy Patterson
Dr Richard Lindzen
Dr Tim Ball
Dr Jordan Peterson
Dr Jonathan Haidt
Dr Stephen Hicks
Dr Eric Weinstein

ARTICLES, BOOKS, AND JOURNALS

Afsah, S. (2013): *Energy Demand Reductions Help Slash US CO2 Emissions: A Closer Analysis.* The Energy Collective. www.theenergycollective.com/shakebafsah/220006/reductio n-energy-demand-helps-slash-us-co2-emissions-analysis

BP (2012). *BP Statistical Review of World Energy June 2012.* www.bp.com/assets/bp_internet/globalbp/globalbp_uk_engl ish/reports_and_publications/statistical_energy_review_201 1/STAGING/local_assets/pdf/statistical_review_of_world_en ergy_full_report_2012.pdf

Ball,T. (2014): *The Deliberate Corruption of Climate Science*

Brignell, J. (2000). *Sorry, Wrong Number; The Abuse of Measurement*. Brignell Associates: London, UK.

CDIAC (2013). *Atmospheric CO2 Records from Sites in the SIO Air Sampling Network*. http://cdiac.ornl.gov/trends/co2/sio-keel.html

Chen,G.Q. Q. Yang, Y.H. Zhao (2011). *Renewability of wind power in China: A case study of nonrenewable energy cost and greenhouse gas emission by a plant in Guangxi*. Review Article. Renewable and Sustainable Energy Reviews, Volume 15, Issue 5, June 2011, Pages 2322-2329.

Cloete, S. (2013). *The Renewable Energy Reality Check*. The Energy Collective. www.theenergycollective.com/schalk-cloete/228151/renewable-energy-reality-check?ref=smt_side_whats_hot

CO2 Now (2013): *www.co2now.org*

DECC (2012): *Annual Energy Statement.*
www.gov.uk/government/publications/annual-energy-statement-2012

Doherty, B. (2006). *Nuclear way to go: Flannery.* The Age.
www.theage.com.au/news/national/nuclear-way-to-go-flannery/2006/08/04/1154198331848.html

Dorling, D. (2013): *Population 10 Billion.*

EIA (2013): *Pumped storage provides grid reliability even with net generation loss.* Today in Energy. US Energy Information Administration.
www.eia.gov/todayinenergy/detail.cfm?id=11991

Energy Development Co-operative Limited (2013):
www.solar-wind.co.uk

EPA (2013a). *Hydroelectricity.* US Environment Protection Agency. www.epa.gov/cleanenergy/energy-and-you/affect/hydro.html

EPRI (2010). *Electricity Energy Storage Technology Options.*
Electric Power Research Institute.
www.epri.com/abstracts/pages/productabstract.aspx?Produ ctID=000000000001020676

European Environment Agency (2013). *Greenhouse gas emission trends (CSI 010) - Assessment published May 2013.*
www.eea.europa.eu/data-and-maps/indicators/greenhouse-gas-emission-trends/greenhouse-gas-emission-trends-assessment-5

Flannery, T. (2005). *The Weather Makers: Our Changing Climate and what it means for Life on Earth.* Penguin Books: London, UK.

Gillespie, D. (2012) Big Fat Lies: How The Diet Industry is Making You Sick, Fat and Poor. Penguin Books: Melbourne, Australia

Goldacre, B. (2009). *Bad Science.* Harper Collins: London, UK.

Gore, A. (2013). *The Future.* Random House: New York, USA.

Gray, J. (2007). *Black Mass; Apocalyptic Religion and the Death of Utopia.* Penguin Books: London, UK.

Greenblatt,J. Samir Succar, David C. Denkenberger, Robert H. Williams, Robert H. Socolow (2007). *Baseload wind energy: modeling the competition between gas turbines and compressed air energy storage for supplemental generation.* Original Research Article. Energy Policy, Volume 35, Issue 3, March 2007, Pages 1474-1492.

Gutiérrez-Martín,F. R.A. Da Silva-Álvarez, P. Montoro-Pintado (2013). *Effects of wind intermittency on reduction of CO2 emissions: The case of the Spanish power system.* Original Research Article. Energy, In Press, Corrected Proof, Available online 4 March 2013.

Guzowski, M. (2010). *Towards Zero Energy Architecture; New Solar Design.* Laurence King Publishing Ltd: London, UK. 55

Harris, C. and Borer, P (1998). *The Whole House Book; Ecological Building Design and Materials.* Centre for Alternative Technology: Machynlleth, UK.

Hausfather, Z. (2013). *What's Behind the 'Good News' Declines in US CO₂ Emissions?*: The Yale Forum on Climate Change and the Media. www.yaleclimatemediaforum.org/2013/05/whats-behind-the-good-news-declines-in-u-s-co2-emissions/

Heinberg, R. (2011). *The End of Growth; Adapting to Our New Economic Reality*. Clairview Books: Forest Row, UK.

Hofmann,D. James H. Butler, Pieter P. Tans (2009). *A new look at atmospheric carbon dioxide.* Atmospheric Environment, Volume 43, Issue 12, April 2009, Pages 2084-2086.

Homewood, P. (2013). *Germany To Open Six More Coal Power Stations In 2013*. Watts Up With That?-The world's most viewed site on global warming and climate change. www.wattsupwiththat.com/2013/04/23/germany-to-open-six-more-coal-power-stations-in-2013/

Institute of Energy Research (2013). *Germany's Green Energy Destabilizing Electric Grids.* www.instituteforenergyresearch.org/2013/01/23/germanys-green-energy-destabilizing-electric-grids/

IRENA: (2013b). *Renewable Power Generation Costs in 2012: An overview*. International Renewable Energy Agency. http://irena.org/DocumentDownloads/Publications/Overview_Renewable%20Power%20Generation%20Costs%20in%202012.pdf

Judt, T. (2011). *Ill Fares the Land*. Penguin Books: London, UK.

Kelly-Detwiler, P (2013). *Denmark: 1,000 Megawatts Of Offshore Wind, And No Signs of Slowing Down.* Forbes. www.forbes.com/sites/peterdetwiler/2013/03/26/denmark-1000-megawatts-of-offshore-wind-and-no-signs-of-slowing-down

Li, X. K. Hubacek, Y.L. Siu. (2012). *Wind power in China – Dream or reality?* Original Research Article. Energy, Volume 37, Issue 1, January 2012, Pages 51-60.

Li,J. (2010). *Decarbonising power generation in China—Is the answer blowing in the wind?* Review Article. Renewable and Sustainable Energy Reviews, Volume 14, Issue 4, May 2010, Pages 1154-1171.

Lomborg, B. (2001). *The Skeptical Environmentalist; Measuring the Real State of the World*. Cambridge University Press: UK.

Lovelock, J. (2007). *The Revenge of Gaia*. Penguin Books: London, UK.

Luoma, J. (2009). *The Challenge for Green Energy: How to Store Excess Electricity*. Environment 360. www.e360.yale.edu/feature/the_challenge_for_green_energy_how_to_store_excess_electricity/2170/

MacKay, D. (2008). *Sustainable Energy – without the hot air*. UIT: Cambridge, UK.

Marohasy, J. (Ed). (2017). *Climate Change: The Facts 2017*.

Martenson, C. (2011). *The Crash Course – The Unsustainable Future of Our Economy, Energy, and Environment*. John Wiley & Sons: New Jersey.

Moore, S (Ed). (2010). *Pragmatic Sustainability; Theoretical and Practical Tools*. Routledge: Abingdon, UK.

Nasr, S. (2009)*: How Grid Energy Storage Works*. How Stuff Works.
http://science.howstuffworks.com/environmental/green-tech/sustainable/grid-energy-storage.htm

National Climatic Data Center. (2013). *State of the Climate: Global Analysis for March 2013*, published online April 2013.
www.ncdc.noaa.gov/sotc/global/

Peltier, R. (2013). *Germany's Expensive Experiment*. Power.
www.powermag.com/germanys-expensive-experiment/

Pinker, S. (2018): *Enlightenment Now; The Case for Reason, Science, Humanism and Progress*.

Plimer, I. (2017): *Climate Change Delusion and the Great Electricity Rip-off*

Porate, K.B. K.L. Thakre, G.L. Bodhe (2007). *Impact of wind power on generation economy and emission from coal based thermal power plant*. Original Research Article. International Journal of Electrical Power & Energy Systems, Volume 44, Issue 1, January 2013, Pages 889-896.

Preston, R. (2012). *How Do We Fix This Mess; The Economic Price of Having it All and the Route to Lasting Prosperity*. Hodder & Stoughton: London, UK.

Rabiee,A. Hossein Khorramdel, Jamshid Aghaei. (2013). *A review of energy storage systems in microgrids with wind turbines*. Review Article. Renewable and Sustainable Energy Reviews, Volume 18, February 2013, Pages 316-326.

Rifkind, H. (6 Aug 2013) The Times p.23.

Rosenbloom, E. (2013). *A Problem With Windpower*.
www.aweo.org/problemwithwind.html

Sandel, M. (2013). *What Money Can't Buy; The Moral Limits of Markets*. Penguin Books: London, UK.

Sassi, P. (2006). *Strategies for Sustainable Architecture*. Taylor and Francis: Abingdon, UK.

Scorah, H. Amy Sopinka, G. Cornelis van Kooten (2011). *The economics of storage, transmission and drought: integrating variable wind power into spatially separated electricity grids*. Original Research Article. Energy Economics, Volume 34, Issue 2, March 2012, Pages 536-541. 57

Scruton, R. (2012). *Green Philosophy; How to Think Seriously About the Planet*. Atlantic Books: London, UK.

Shafiullah,GM. Amanullah M.T. Oo, A.B.M. Shawkat Ali, Peter Wolfs (2013). *Potential challenges of integrating large-scale wind energy into the power grid–A review*. Review Article. Renewable and Sustainable Energy Reviews, Volume 20, April 2013, Pages 306-321.

Sharman, R. (2013). Tasman Energy.
www.tasmanenergy.com.au

Sheldrake, R. (2013). *The Science Delusion*. Hodder & Stoughton: London, UK.

Smart Grid (2013). US Department of Energy. www.smartgrid.gov/the_smart_grid#smart_grid

Solar Choice (2013). www.solarchoice.net.au

Taleb, N. (2007). *The Black Swan; The Impact of the Highly Improbable*. Penguin Books: London, UK.

The Eco Experts (2013): www.theecoexperts.co.uk

Wai-yin Kwok, V (2009): *Weaknesses In Chinese Wind Power*. Forbes. www.forbes.com/2009/07/20/china-wind-power-business-energy-china.html

Yergin, Y. (2011). *The Quest; Energy Security, and the Remaking of the Modern World*. Penguin Books: London, UK.